T0137031

Lecture Notes in Networks and Systems   985

The series "Lecture Notes in Networks and Systems" publishes the latest developments in Networks and Systems—quickly, informally and with high quality. Original research reported in proceedings and post-proceedings represents the core of LNNS.

Volumes published in LNNS embrace all aspects and subfields of, as well as new challenges in, Networks and Systems.

The series contains proceedings and edited volumes in systems and networks, spanning the areas of Cyber-Physical Systems, Autonomous Systems, Sensor Networks, Control Systems, Energy Systems, Automotive Systems, Biological Systems, Vehicular Networking and Connected Vehicles, Aerospace Systems, Automation, Manufacturing, Smart Grids, Nonlinear Systems, Power Systems, Robotics, Social Systems, Economic Systems and other. Of particular value to both the contributors and the readership are the short publication timeframe and the worldwide distribution and exposure which enable both a wide and rapid dissemination of research output.

The series covers the theory, applications, and perspectives on the state of the art and future developments relevant to systems and networks, decision making, control, complex processes and related areas, as embedded in the fields of interdisciplinary and applied sciences, engineering, computer science, physics, economics, social, and life sciences, as well as the paradigms and methodologies behind them.

Indexed by SCOPUS, INSPEC, WTI Frankfurt eG, zbMATH, SCImago.

All books published in the series are submitted for consideration in Web of Science.

For proposals from Asia please contact Aninda Bose (aninda.bose@springer.com).

Álvaro Rocha · Hojjat Adeli ·
Gintautas Dzemyda · Fernando Moreira ·
Aneta Poniszewska-Marańda
Editors

Good Practices and New Perspectives in Information Systems and Technologies

WorldCIST 2024, Volume 1

 Springer

Editors
Álvaro Rocha
ISEG
Universidade de Lisboa
Lisbon, Portugal

Hojjat Adeli
College of Engineering
The Ohio State University
Columbus, OH, USA

Gintautas Dzemyda
Institute of Data Science and Digital
Technologies
Vilnius University
Vilnius, Lithuania

Fernando Moreira
DCT
Universidade Portucalense
Porto, Portugal

Aneta Poniszewska-Marańda
Institute of Information Technology
Lodz University of Technology
Łódz, Poland

ISSN 2367-3370   ISSN 2367-3389 (electronic)
Lecture Notes in Networks and Systems
ISBN 978-3-031-60214-6   ISBN 978-3-031-60215-3 (eBook)
https://doi.org/10.1007/978-3-031-60215-3

This Springer imprint is published by the registered company Springer Nature Switzerland AG
The registered company address is: Gewerbestrasse 11, 6330 Cham, Switzerland

Preface

This book contains a selection of papers accepted for presentation and discussion at the 2024 World Conference on Information Systems and Technologies (WorldCIST'24). This conference had the scientific support of the Lodz University of Technology, Information and Technology Management Association (ITMA), IEEE Systems, Man, and Cybernetics Society (IEEE SMC), Iberian Association for Information Systems and Technologies (AISTI), and Global Institute for IT Management (GIIM). It took place in Lodz city, Poland, 26–28 March 2024.

The World Conference on Information Systems and Technologies (WorldCIST) is a global forum for researchers and practitioners to present and discuss recent results and innovations, current trends, professional experiences, and challenges of modern Information Systems and Technologies research, technological development, and applications. One of its main aims is to strengthen the drive toward a holistic symbiosis between academy, society, and industry. WorldCIST'23 is built on the successes of: WorldCIST'13 held at Olhão, Algarve, Portugal; WorldCIST'14 held at Funchal, Madeira, Portugal; WorldCIST'15 held at São Miguel, Azores, Portugal; WorldCIST'16 held at Recife, Pernambuco, Brazil; WorldCIST'17 held at Porto Santo, Madeira, Portugal; WorldCIST'18 held at Naples, Italy; WorldCIST'19 held at La Toja, Spain; WorldCIST'20 held at Budva, Montenegro; WorldCIST'21 held at Terceira Island, Portugal; WorldCIST'22 held at Budva, Montenegro; and WorldCIST'23, which took place at Pisa, Italy.

The Program Committee of WorldCIST'24 was composed of a multidisciplinary group of 328 experts and those who are intimately concerned with Information Systems and Technologies. They have had the responsibility for evaluating, in a 'blind review' process, the papers received for each of the main themes proposed for the conference: A) Information and Knowledge Management; B) Organizational Models and Information Systems; C) Software and Systems Modeling; D) Software Systems, Architectures, Applications and Tools; E) Multimedia Systems and Applications; F) Computer Networks, Mobility and Pervasive Systems; G) Intelligent and Decision Support Systems; H) Big Data Analytics and Applications; I) Human-Computer Interaction; J) Ethics, Computers & Security; K) Health Informatics; L) Information Technologies in Education; M) Information Technologies in Radiocommunications; and N) Technologies for Biomedical Applications.

The conference also included workshop sessions taking place in parallel with the conference ones. Workshop sessions covered themes such as: ICT for Auditing & Accounting; Open Learning and Inclusive Education Through Information and Communication Technology; Digital Marketing and Communication, Technologies, and Applications; Advances in Deep Learning Methods and Evolutionary Computing for Health Care; Data Mining and Machine Learning in Smart Cities: The role of the technologies in the research of the migrations; Artificial Intelligence Models and Artifacts for Business Intelligence Applications; AI in Education; Environmental data analytics; Forest-Inspired

Computational Intelligence Methods and Applications; Railway Operations, Modeling and Safety; Technology Management in the Electrical Generation Industry: Capacity Building through Knowledge, Resources and Networks; Data Privacy and Protection in Modern Technologies; Strategies and Challenges in Modern NLP: From Argumentation to Ethical Deployment; and Enabling Software Engineering Practices Via Last Development Trends.

WorldCIST'24 and its workshops received about 400 contributions from 47 countries around the world. The papers accepted for oral presentation and discussion at the conference are published by Springer (this book) in four volumes and will be submitted for indexing by WoS, Scopus, EI-Compendex, DBLP, and/or Google Scholar, among others. Extended versions of selected best papers will be published in special or regular issues of leading and relevant journals, mainly JCR/SCI/SSCI and Scopus/EI-Compendex indexed journals.

We acknowledge all of those that contributed to the staging of WorldCIST'24 (authors, committees, workshop organizers, and sponsors). We deeply appreciate their involvement and support that was crucial for the success of WorldCIST'24.

March 2024

Álvaro Rocha
Hojjat Adeli
Gintautas Dzemyda
Fernando Moreira
Aneta Poniszewska-Marańda

Organization

Conference

Honorary Chair

Hojjat Adeli   The Ohio State University, USA

General Chair

Álvaro Rocha   ISEG, University of Lisbon, Portugal

Co-chairs

Gintautas Dzemyda   Vilnius University, Lithuania
Sandra Costanzo   University of Calabria, Italy

Workshops Chair

Fernando Moreira   Portucalense University, Portugal

Local Organizing Committee

Bożena Borowska   Lodz University of Technology, Poland
Łukasz Chomątek   Lodz University of Technology, Poland
Joanna Ochelska-Mierzejewska   Lodz University of Technology, Poland
Aneta Poniszewska-Marańda   Lodz University of Technology, Poland

Advisory Committee

Ana Maria Correia (Chair)   University of Sheffield, UK
Brandon Randolph-Seng   Texas A&M University, USA

Chris Kimble	KEDGE Business School & MRM, UM2, Montpellier, France
Damian Niwiński	University of Warsaw, Poland
Eugene Spafford	Purdue University, USA
Florin Gheorghe Filip	Romanian Academy, Romania
Janusz Kacprzyk	Polish Academy of Sciences, Poland
João Tavares	University of Porto, Portugal
Jon Hall	The Open University, UK
John MacIntyre	University of Sunderland, UK
Karl Stroetmann	Empirica Communication & Technology Research, Germany
Marjan Mernik	University of Maribor, Slovenia
Miguel-Angel Sicilia	University of Alcalá, Spain
Mirjana Ivanovic	University of Novi Sad, Serbia
Paulo Novais	University of Minho, Portugal
Sami Habib	Kuwait University, Kuwait
Wim Van Grembergen	University of Antwerp, Belgium

Program Committee Co-chairs

Adam Wojciechowski	Lodz University of Technology, Poland
Aneta Poniszewska-Marańda	Lodz University of Technology, Poland

Program Committee

Abderrahmane Ez-zahout	Mohammed V University, Morocco
Adriana Peña Pérez Negrón	Universidad de Guadalajara, Mexico
Adriani Besimi	South East European University, North Macedonia
Agostinho Sousa Pinto	Polytechnic of Porto, Portugal
Ahmed El Oualkadi	Abdelmalek Essaadi University, Morocco
Akex Rabasa	University Miguel Hernandez, Spain
Alanio de Lima	UFC, Brazil
Alba Córdoba-Cabús	University of Malaga, Spain
Alberto Freitas	FMUP, University of Porto, Portugal
Aleksandra Labus	University of Belgrade, Serbia
Alessio De Santo	HE-ARC, Switzerland
Alexandru Vulpe	University Politechnica of Bucharest, Romania
Ali Idri	ENSIAS, University Mohamed V, Morocco
Alicia García-Holgado	University of Salamanca, Spain

Almir Souza Silva Neto	IFMA, Brazil
Álvaro López-Martín	University of Malaga, Spain
Amélia Badica	Universiti of Craiova, Romania
Amélia Cristina Ferreira Silva	Polytechnic of Porto, Portugal
Amit Shelef	Sapir Academic College, Israel
Ana Carla Amaro	Universidade de Aveiro, Portugal
Ana Dinis	Polytechnic of Cávado and Ave, Portugal
Ana Isabel Martins	University of Aveiro, Portugal
Anabela Gomes	University of Coimbra, Portugal
Anacleto Correia	CINAV, Portugal
Andrew Brosnan	University College Cork, Ireland
Andjela Draganic	University of Montenegro, Montenegro
Aneta Polewko-Klim	University of Białystok, Institute of Informatics, Poland
Aneta Poniszewska-Maranda	Lodz University of Technology, Poland
Angeles Quezada	Instituto Tecnologico de Tijuana, Mexico
Anis Tissaoui	University of Jendouba, Tunisia
Ankur Singh Bist	KIET, India
Ann Svensson	University West, Sweden
Anna Gawrońska	Poznański Instytut Technologiczny, Poland
Antoni Oliver	University of the Balearic Islands, Spain
Antonio Jiménez-Martín	Universidad Politécnica de Madrid, Spain
Aroon Abbu	Bell and Howell, USA
Arslan Enikeev	Kazan Federal University, Russia
Beatriz Berrios Aguayo	University of Jaen, Spain
Benedita Malheiro	Polytechnic of Porto, ISEP, Portugal
Bertil Marques	Polytechnic of Porto, ISEP, Portugal
Boris Shishkov	ULSIT/IMI - BAS/IICREST, Bulgaria
Borja Bordel	Universidad Politécnica de Madrid, Spain
Branko Perisic	Faculty of Technical Sciences, Serbia
Bruno F. Gonçalves	Polytechnic of Bragança, Portugal
Carla Pinto	Polytechnic of Porto, ISEP, Portugal
Carlos Balsa	Polytechnic of Bragança, Portugal
Carlos Rompante Cunha	Polytechnic of Bragança, Portugal
Catarina Reis	Polytechnic of Leiria, Portugal
Célio Gonçalo Marques	Polytenic of Tomar, Portugal
Cengiz Acarturk	Middle East Technical University, Turkey
Cesar Collazos	Universidad del Cauca, Colombia
Cristina Gois	Polytechnic University of Coimbra, Portugal
Christophe Guyeux	Universite de Bourgogne Franche Comté, France
Christophe Soares	University Fernando Pessoa, Portugal
Christos Bouras	University of Patras, Greece

Christos Chrysoulas	London South Bank University, UK
Christos Chrysoulas	Edinburgh Napier University, UK
Ciro Martins	University of Aveiro, Portugal
Claudio Sapateiro	Polytechnic of Setúbal, Portugal
Cosmin Striletchi	Technical University of Cluj-Napoca, Romania
Costin Badica	University of Craiova, Romania
Cristian García Bauza	PLADEMA-UNICEN-CONICET, Argentina
Cristina Caridade	Polytechnic of Coimbra, Portugal
Danish Jamil	Malaysia University of Science and Technology, Malaysia
David Cortés-Polo	University of Extremadura, Spain
David Kelly	University College London, UK
Daria Bylieva	Peter the Great St. Petersburg Polytechnic University, Russia
Dayana Spagnuelo	Vrije Universiteit Amsterdam, Netherlands
Dhouha Jaziri	University of Sousse, Tunisia
Dmitry Frolov	HSE University, Russia
Dulce Mourato	ISTEC - Higher Advanced Technologies Institute Lisbon, Portugal
Edita Butrime	Lithuanian University of Health Sciences, Lithuania
Edna Dias Canedo	University of Brasilia, Brazil
Egils Ginters	Riga Technical University, Latvia
Ekaterina Isaeva	Perm State University, Russia
Eliana Leite	University of Minho, Portugal
Enrique Pelaez	ESPOL University, Ecuador
Eriks Sneiders	Stockholm University, Sweden; Esteban Castellanos ESPE, Ecuador
Fatima Azzahra Amazal	Ibn Zohr University, Morocco
Fernando Bobillo	University of Zaragoza, Spain
Fernando Molina-Granja	National University of Chimborazo, Ecuador
Fernando Moreira	Portucalense University, Portugal
Fernando Ribeiro	Polytechnic Castelo Branco, Portugal
Filipe Caldeira	Polytechnic of Viseu, Portugal
Filippo Neri	University of Naples, Italy
Firat Bestepe	Republic of Turkey Ministry of Development, Turkey
Francesco Bianconi	Università degli Studi di Perugia, Italy
Francisco García-Peñalvo	University of Salamanca, Spain
Francisco Valverde	Universidad Central del Ecuador, Ecuador
Frederico Branco	University of Trás-os-Montes e Alto Douro, Portugal
Galim Vakhitov	Kazan Federal University, Russia

Gayo Diallo	University of Bordeaux, France
Gabriel Pestana	Polytechnic Institute of Setubal, Portugal
Gema Bello-Orgaz	Universidad Politecnica de Madrid, Spain
George Suciu	BEIA Consult International, Romania
Ghani Albaali	Princess Sumaya University for Technology, Jordan
Gian Piero Zarri	University Paris-Sorbonne, France
Giovanni Buonanno	University of Calabria, Italy
Gonçalo Paiva Dias	University of Aveiro, Portugal
Goreti Marreiros	ISEP/GECAD, Portugal
Habiba Drias	University of Science and Technology Houari Boumediene, Algeria
Hafed Zarzour	University of Souk Ahras, Algeria
Haji Gul	City University of Science and Information Technology, Pakistan
Hakima Benali Mellah	Cerist, Algeria
Hamid Alasadi	Basra University, Iraq
Hatem Ben Sta	University of Tunis at El Manar, Tunisia
Hector Fernando Gomez Alvarado	Universidad Tecnica de Ambato, Ecuador
Hector Menendez	King's College London, UK
Hélder Gomes	University of Aveiro, Portugal
Helia Guerra	University of the Azores, Portugal
Henrique da Mota Silveira	University of Campinas (UNICAMP), Brazil
Henrique S. Mamede	University Aberta, Portugal
Henrique Vicente	University of Évora, Portugal
Hicham Gueddah	University Mohammed V in Rabat, Morocco
Hing Kai Chan	University of Nottingham Ningbo China, China
Igor Aguilar Alonso	Universidad Nacional Tecnológica de Lima Sur, Peru
Inês Domingues	University of Coimbra, Portugal
Isabel Lopes	Polytechnic of Bragança, Portugal
Isabel Pedrosa	Coimbra Business School - ISCAC, Portugal
Isaías Martins	University of Leon, Spain
Issam Moghrabi	Gulf University for Science and Technology, Kuwait
Ivan Armuelles Voinov	University of Panama, Panama
Ivan Dunđer	University of Zagreb, Croatia
Ivone Amorim	University of Porto, Portugal
Jaime Diaz	University of La Frontera, Chile
Jan Egger	IKIM, Germany
Jan Kubicek	Technical University of Ostrava, Czech Republic
Jeimi Cano	Universidad de los Andes, Colombia

Jesús Gallardo Casero	University of Zaragoza, Spain
Jezreel Mejia	CIMAT, Unidad Zacatecas, Mexico
Jikai Li	The College of New Jersey, USA
Jinzhi Lu	KTH-Royal Institute of Technology, Sweden
Joao Carlos Silva	IPCA, Portugal
João Manuel R. S. Tavares	University of Porto, FEUP, Portugal
João Paulo Pereira	Polytechnic of Bragança, Portugal
João Reis	University of Aveiro, Portugal
João Reis	University of Lisbon, Portugal
João Rodrigues	University of the Algarve, Portugal
João Vidal de Carvalho	Polytechnic of Porto, Portugal
Joaquin Nicolas Ros	University of Murcia, Spain
John W. Castro	University de Atacama, Chile
Jorge Barbosa	Polytechnic of Coimbra, Portugal
Jorge Buele	Technical University of Ambato, Ecuador; Jorge Gomes University of Lisbon, Portugal
Jorge Oliveira e Sá	University of Minho, Portugal
José Braga de Vasconcelos	Universidade Lusófona, Portugal
Jose M. Parente de Oliveira	Aeronautics Institute of Technology, Brazil
José Machado	University of Minho, Portugal
José Paulo Lousado	Polytechnic of Viseu, Portugal
Josc Quiroga	University of Oviedo, Spain
Jose Silvestre Silva	Academia Military, Portugal
Jose Torres	University Fernando Pessoa, Portugal
Juan M. Santos	University of Vigo, Spain
Juan Manuel Carrillo de Gea	University of Murcia, Spain
Juan Pablo Damato	UNCPBA-CONICET, Argentina
Kalinka Kaloyanova	Sofia University, Bulgaria
Kamran Shaukat	The University of Newcastle, Australia
Katerina Zdravkova	University Ss. Cyril and Methodius, North Macedonia
Khawla Tadist	Morocco
Khalid Benali	LORIA - University of Lorraine, France
Khalid Nafil	Mohammed V University in Rabat, Morocco
Korhan Gunel	Adnan Menderes University, Turkey
Krzysztof Wolk	Polish-Japanese Academy of Information Technology, Poland
Kuan Yew Wong	Universiti Teknologi Malaysia (UTM), Malaysia
Kwanghoon Kim	Kyonggi University, South Korea
Laila Cheikhi	Mohammed V University in Rabat, Morocco
Laura Varela-Candamio	Universidade da Coruña, Spain
Laurentiu Boicescu	E.T.T.I. U.P.B., Romania

Lbtissam Abnane	ENSIAS, Morocco
Lia-Anca Hangan	Technical University of Cluj-Napoca, Romania
Ligia Martinez	CECAR, Colombia
Lila Rao-Graham	University of the West Indies, Jamaica
Liliana Ivone Pereira	Polytechnic of Cávado and Ave, Portugal
Łukasz Tomczyk	Pedagogical University of Cracow, Poland
Luis Alvarez Sabucedo	University of Vigo, Spain
Luís Filipe Barbosa	University of Trás-os-Montes e Alto Douro
Luis Mendes Gomes	University of the Azores, Portugal
Luis Pinto Ferreira	Polytechnic of Porto, Portugal
Luis Roseiro	Polytechnic of Coimbra, Portugal
Luis Silva Rodrigues	Polytencic of Porto, Portugal
Mahdieh Zakizadeh	MOP, Iran
Maksim Goman	JKU, Austria
Manal el Bajta	ENSIAS, Morocco
Manuel Antonio Fernández-Villacañas Marín	Technical University of Madrid, Spain
Manuel Ignacio Ayala Chauvin	University Indoamerica, Ecuador
Manuel Silva	Polytechnic of Porto and INESC TEC, Portugal
Manuel Tupia	Pontifical Catholic University of Peru, Peru
Manuel Au-Yong-Oliveira	University of Aveiro, Portugal
Marcelo Mendonça Teixeira	Universidade de Pernambuco, Brazil
Marciele Bernardes	University of Minho, Brazil
Marco Ronchetti	Universita' di Trento, Italy
Mareca María Pilar	Universidad Politécnica de Madrid, Spain
Marek Kvet	Zilinska Univerzita v Ziline, Slovakia
Maria João Ferreira	Universidade Portucalense, Portugal
Maria José Sousa	University of Coimbra, Portugal
María Teresa García-Álvarez	University of A Coruna, Spain
Maria Sokhn	University of Applied Sciences of Western Switzerland, Switzerland
Marijana Despotovic-Zrakic	Faculty Organizational Science, Serbia
Marilio Cardoso	Polytechnic of Porto, Portugal
Mário Antunes	Polytechnic of Leiria & CRACS INESC TEC, Portugal
Marisa Maximiano	Polytechnic Institute of Leiria, Portugal
Marisol Garcia-Valls	Polytechnic University of Valencia, Spain
Maristela Holanda	University of Brasilia, Brazil
Marius Vochin	E.T.T.I. U.P.B., Romania
Martin Henkel	Stockholm University, Sweden
Martín López Nores	University of Vigo, Spain
Martin Zelm	INTEROP-VLab, Belgium

Mazyar Zand	MOP, Iran
Mawloud Mosbah	University 20 Août 1955 of Skikda, Algeria
Michal Adamczak	Poznan School of Logistics, Poland
Michal Kvet	University of Zilina, Slovakia
Miguel Garcia	University of Oviedo, Spain
Mircea Georgescu	Al. I. Cuza University of Iasi, Romania
Mirna Muñoz	Centro de Investigación en Matemáticas A.C., Mexico
Mohamed Hosni	ENSIAS, Morocco
Monica Leba	University of Petrosani, Romania
Nadesda Abbas	UBO, Chile
Narasimha Rao Vajjhala	University of New York Tirana, Tirana
Narjes Benameur	Laboratory of Biophysics and Medical Technologies of Tunis, Tunisia
Natalia Grafeeva	Saint Petersburg University, Russia
Natalia Miloslavskaya	National Research Nuclear University MEPhI, Russia
Naveed Ahmed	University of Sharjah, United Arab Emirates
Neeraj Gupta	KIET group of institutions Ghaziabad, India
Nelson Rocha	University of Aveiro, Portugal
Nikola S. Nikolov	University of Limerick, Ireland
Nicolas de Araujo Moreira	Federal University of Ceara, Brazil
Nikolai Prokopyev	Kazan Federal University, Russia
Niranjan S. K.	JSS Science and Technology University, India
Noemi Emanuela Cazzaniga	Politecnico di Milano, Italy
Noureddine Kerzazi	Polytechnique Montréal, Canada
Nuno Melão	Polytechnic of Viseu, Portugal
Nuno Octávio Fernandes	Polytechnic of Castelo Branco, Portugal
Nuno Pombo	University of Beira Interior, Portugal
Olga Kurasova	Vilnius University, Lithuania
Olimpiu Stoicuta	University of Petrosani, Romania
Patricia Quesado	Polytechnic of Cávado and Ave, Portugal
Patricia Zachman	Universidad Nacional del Chaco Austral, Argentina
Paula Serdeira Azevedo	University of Algarve, Portugal
Paula Dias	Polytechnic of Guarda, Portugal
Paulo Alejandro Quezada Sarmiento	University of the Basque Country, Spain
Paulo Maio	Polytechnic of Porto, ISEP, Portugal
Paulvanna Nayaki Marimuthu	Kuwait University, Kuwait
Paweł Karczmarek	The John Paul II Catholic University of Lublin, Poland

Pedro Rangel Henriques	University of Minho, Portugal
Pedro Sobral	University Fernando Pessoa, Portugal
Pedro Sousa	University of Minho, Portugal
Philipp Jordan	University of Hawaii at Manoa, USA
Piotr Kulczycki	Systems Research Institute, Polish Academy of Sciences, Poland
Prabhat Mahanti	University of New Brunswick, Canada
Rabia Azzi	Bordeaux University, France
Radu-Emil Precup	Politehnica University of Timisoara, Romania
Rafael Caldeirinha	Polytechnic of Leiria, Portugal
Raghuraman Rangarajan	Sequoia AT, Portugal
Radhakrishna Bhat	Manipal Institute of Technology, India
Raiani Ali	Hamad Bin Khalifa University, Qatar
Ramadan Elaiess	University of Benghazi, Libya
Ramayah T.	Universiti Sains Malaysia, Malaysia
Ramazy Mahmoudi	University of Monastir, Tunisia
Ramiro Gonçalves	University of Trás-os-Montes e Alto Douro & INESC TEC, Portugal
Ramon Alcarria	Universidad Politécnica de Madrid, Spain
Ramon Fabregat Gesa	University of Girona, Spain
Ramy Rahimi	Chungnam National University, South Korea
Reiko Hishiyama	Waseda University, Japan
Renata Maria Maracho	Federal University of Minas Gerais, Brazil
Renato Toasa	Israel Technological University, Ecuador
Reyes Juárez Ramírez	Universidad Autonoma de Baja California, Mexico
Rocío González-Sánchez	Rey Juan Carlos University, Spain
Rodrigo Franklin Frogeri	University Center of Minas Gerais South, Brazil
Ruben Pereira	ISCTE, Portugal
Rui Alexandre Castanho	WSB University, Poland
Rui S. Moreira	UFP & INESC TEC & LIACC, Portugal
Rustam Burnashev	Kazan Federal University, Russia
Saeed Salah	Al-Quds University, Palestine
Said Achchab	Mohammed V University in Rabat, Morocco
Sajid Anwar	Institute of Management Sciences Peshawar, Pakistan
Sami Habib	Kuwait University, Kuwait
Samuel Sepulveda	University of La Frontera, Chile
Sara Luis Dias	Polytechnic of Cávado and Ave, Portugal
Sandra Costanzo	University of Calabria, Italy
Sandra Patricia Cano Mazuera	University of San Buenaventura Cali, Colombia
Sassi Sassi	FSJEGJ, Tunisia

Seppo Sirkemaa	University of Turku, Finland
Sergio Correia	Polytechnic of Portalegre, Portugal
Shahnawaz Talpur	Mehran University of Engineering & Technology Jamshoro, Pakistan
Shakti Kundu	Manipal University Jaipur, Rajasthan, India
Shashi Kant Gupta	Eudoxia Research University, USA
Silviu Vert	Politehnica University of Timisoara, Romania
Simona Mirela Riurean	University of Petrosani, Romania
Slawomir Zolkiewski	Silesian University of Technology, Poland
Solange Rito Lima	University of Minho, Portugal
Sonia Morgado	ISCPSI, Portugal
Sonia Sobral	Portucalense University, Portugal
Sorin Zoican	Polytechnic University of Bucharest, Romania
Souraya Hamida	Batna 2 University, Algeria
Stalin Figueroa	University of Alcala, Spain
Sümeyya Ilkin	Kocaeli University, Turkey
Syed Asim Ali	University of Karachi, Pakistan
Syed Nasirin	Universiti Malaysia Sabah, Malaysia
Tatiana Antipova	Institute of Certified Specialists, Russia
TatiannaRosal	University of Trás-os-Montes e Alto Douro, Portugal
Tero Kokkonen	JAMK University of Applied Sciences, Finland
The Thanh Van	HCMC University of Food Industry, Vietnam
Thomas Weber	EPFL, Switzerland
Timothy Asiedu	TIM Technology Services Ltd., Ghana
Tom Sander	New College of Humanities, Germany
Tomasz Kisielewicz	Warsaw University of Technology
Tomaž Klobučar	Jozef Stefan Institute, Slovenia
Toshihiko Kato	University of Electro-communications, Japan
Tuomo Sipola	Jamk University of Applied Sciences, Finland
Tzung-Pei Hong	National University of Kaohsiung, Taiwan
Valentim Realinho	Polytechnic of Portalegre, Portugal
Valentina Colla	Scuola Superiore Sant'Anna, Italy
Valerio Stallone	ZHAW, Switzerland
Verónica Vasconcelos	Polytechnic of Coimbra, Portugal
Vicenzo Iannino	Scuola Superiore Sant'Anna, Italy
Vitor Gonçalves	Polytechnic of Bragança, Portugal
Victor Alves	University of Minho, Portugal
Victor Georgiev	Kazan Federal University, Russia
Victor Hugo Medina Garcia	Universidad Distrital Francisco José de Caldas, Colombia
Victor Kaptelinin	Umeå University, Sweden

Viktor Medvedev	Vilnius University, Lithuania
Vincenza Carchiolo	University of Catania, Italy
Waqas Bangyal	University of Gujrat, Pakistan
Wolf Zimmermann	Martin Luther University Halle-Wittenberg, Germany
Yadira Quiñonez	Autonomous University of Sinaloa, Mexico
Yair Wiseman	Bar-Ilan University, Israel
Yassine Drias	University of Algiers, Algeria
Yuhua Li	Cardiff University, UK
Yuwei Lin	University of Roehampton, UK
Zbigniew Suraj	University of Rzeszow, Poland
Zorica Bogdanovic	University of Belgrade, Serbia

Contents

Ethics, Computers and Security

Human-Computer Interaction

Big Data Analytics and Applications

An Overview on the Use of Machine Learning Algorithms for Identifying Anomalies in Industrial Valves

Lesly Ttito Ugarte$^{(\boxtimes)}$ (iD) and Flavia Bernardini (iD)

Instituto de Computação, Universidade Federal Fluminense, Niterói, Brazil
leslystu@id.uff.br, fcbernardini@ic.uff.br

Abstract. Valves used in industrial systems may present anomalous behaviour or eventually fail throughout their useful life. The automatic detection of these anomalies and failures is quite important nowadays, specially due to the increasing movement of industry 4.0 and Digital Twins technologies. In order to detect their anomalies or failures, data captured by sensors in these valves can be used for analysis and prediction tasks. More specifically, these time series data are analyzed and used for constructing machine-learning based models for predicting, which may contribute to better preventive maintenance and longer valve life. The objective of this work is to present the results obtained when executing a Systematic Literature Review (SLR) in order to identify the state of the art on using Machine Learning (ML) algorithms and methods to identify anomalies in industrial valves. Case studies could be identified on different types of valves, including hydraulic, solenoid, gas, nuclear and compression valves. This indicates a long future to develop and propose solutions in this field of science. An increasing focus on the use of deep learning models could also be observed, although there are also simpler methods under scrutiny. Finally, it could be observed that it is quite difficult to find available datasets in order to replicate the obtained results, as well as advance in new ideas and methods in this area.

Keywords: anomaly detection · industry 4.0 · digital twins · machine learning · time series analysis

1 Introduction

According to Morteza [5], the fourth industrial revolution, known as Industry 4.0, is progressively present in the industrial processes, which is mainly focused on digital transformation. Also, the author states that one important technology framework used in this scenario is regarded to the big data analytics, aiming to create a dynamic cyber-physical control system for improving efficiency and reliability in industrial operations. In this Industry 4.0 scenario, emerged the Digital Twin concept, which also aims to offer to industrial processes managers and operators a cyber-physical system as cost-effective solutions to meet stakeholder requirements, including the analysis and simulation of operations [7]. In

Á. Rocha et al. (Eds.): WorldCIST 2024, LNNS 985, pp. 3–12, 2024.
https://doi.org/10.1007/978-3-031-60215-3_1

the context of industrial operations using valves, such as hydraulic, solenoid, gas, nuclear and compression valves. Predicting their Remaining Useful Life (RUL) or anomalous behaviour are important tasks in industrial systems [2]. These tasks can be benefited from the Digital Twins in Industry 4.0 scenario, as they have data collected and stored from sensors.

Valve state detection and RUL prediction methods of industrial valves include physics-based and data-driven approaches [13]. Physics-based methods rely on the deviation between the system and the physical model to detect the operating state of the system. The data-driven approach establishes the relationship between system states and system measurements by analyzing previously collected information about the system. The advantages of data-driven methods are that sufficient prior knowledge is not necessary and an accurate physical model does not need to be established, status diagnosis can be performed by analyzing changes in status signals [20], Data-based methods have several submethods within which are composed by ML algorithms [13].

The objective of this work is to present the results of a SLR, conducted for identifying the state of the art on the use of methodologies, methods and techniques to identify anomalies in industrial valves. The main purpose in this study is to answer the Research Question (RQ): Which machine learning-based algorithms and methods are used to identify anomalies in industrial valves? More specifically, the PRISMA methodology [14] was used for conducting the SLR. An increasing focus could be seen on the use of deep learning-based models, although there are also simpler methods under scrutiny. Finally, could be observed that may be quite difficult to find available datasets for replicating the obtained results and advancing the state-of-the-art in this area.

This work is organized as follows: Sect. 2 presents the methodology for conducting the SLR. Section 3 presents a summary of the findings. Section 4 presents a brief description of the selected studies and a discussion on the datasets publicly available for reuse and replicability. Finally, Sect. 5 presents the conclusions as well as limitations and future work.

2   Methodology for Conducting the SLR

For conducting the SRL, executed according to the PRISMA methodology [14], The following digital libraries were used: ACM Digital Library, IEEE Digital Library and Scopus. They were selected due to be the ones that index the majority of works in computer science and engineering. The RQ was: Which machine learning-based algorithms and methods are used to identify anomalies in industrial valves? Works were collected using the search string ("anomaly" OR "failure" OR "prediction") AND ("valve") AND ("machine learning" OR "deep learning" OR "genetic algorithm" OR "neural networks"), adapted for each digital library.

A total of 191 works were collected, with publication years between 2018 and 2023, being 96 from Scopus; 70 from ACM Digital Library; and 25 from IEEE. An Inclusion Criteria (IC) was applied: (IC1) The work focuses on identifying anomalies or failures on some type of industrial valve and using ML methods. It was also used the following Exclusion Criteria (EC): (EC1) duplicate works; (EC2) secondary studies; (EC3) other types of valve anomalies such as chemical analysis of substances flowing through valves, biomechanical valves or valves inside engines. This is due to the main interest was related to underwater valves or valves used for measuring pressure, flow rate, and others. The main goal was to understand the types of methods the literature are proposing to these types of valves. Aplying the IC and EC, finally, 23 of 191 works were selected. Figure 1 shows a flow chart describing the number of identified, excluded and included works during the SLR phases.

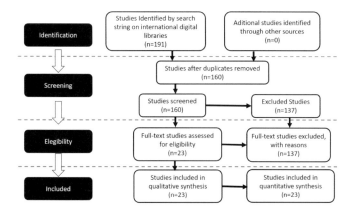

Fig. 1. Information flow through the SLR steps.

3   Summary of the Findings

Analyzing the 23 studies that were selected, only 4 of them made a comparison between the different ML algorithms they decided to use. Also, 10 out of 23 relied on general ML methods. Finally, 16 out of 23 trained Artificial Neural Networks (ANN) models. From the perspective of the authors, Deep Neural Networks (DNN), including Long Short-Term Memory (LSTM) and Convolutional Neural Networks (CNN), are specific structures and types of ANNs. The other types of ANN that are not DNN are called Non-Deep Neural Networks (Non-Deep-NN), which includes MultiLayer Perceptron (MLP), Radial-Basis Function Network (RBF), among others. So, The works that use ANN were divided into three categories.: (i) Non-Deep-NN-based methods; (ii) LSTM-based methods; and (iii) CNN-based methods. Table 1 shows which groups of ML algorithms are used by the different works that were selected and studied.. Two Venn diagrams were also constructed to relate the different types of ML algorithms that could be found in the selected works. Figure 2 shows these diagrams with the number of studies by type and those that used both deep learning and ML methods and a comparison between them. Among the studies that decided to use ANN-based Methods, 9 decided to use Non-Deep NN-based methods, 5 used LSTM-based methods, and 6 decided to use CNN-based methods. The intersections show works in which more than one ANN-Based Method algorithm was used.

Table 1. List of papers and the ML algorithms they used

ML-based Methods	Studies	Total
Non-Deep-NN-based methods	$[1,2,8\text{--}10,13,15,17,22]$	9
CNN-based Methods	$[6,12,19,20,23,25]$	6
LSTM-based Methods	$[1,4,13,19,23]$	5
SVM	$[3,9,11,22]$	4
DT-based Methods	$[2,3,9,11,16,18,21]$	7
Random Forest	$[9,16,21]$	3
KNN	$[3,21],$	2
Logistic regression	$[9,21]$	2
Naive Bayes	$[21]$	1

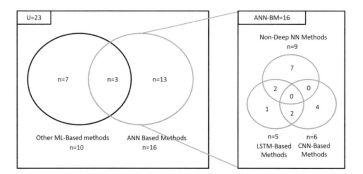

Fig. 2. Number of studies by type of algorithms used.

4   Deeper Analysis of the Findings

Supervised algorithms need humans to give input and required output, in addition to providing feedback about the prediction accuracy in the training process. Within the studies found, it was found that supervised learning algorithms are the most used. No studies were found that used unsupervised learning and only one study using semi-supervised learning, which may mean that is more common to find (partially) labeled than unlabeled data in this scenario.

First, a summary of the selected works that propose methods is presented **based on Non-Deep-NN**: **(i) Andrade et al.** [2] propose a model to diagnose faults through isolation using residual patterns and the construction of a Decision Tree (DT). It uses Fault Detection and Diagnosis (FDD) techniques, through the Nonlinear Auto-Regressive Neural Network Model with Exogenous Inputs (NARX), in the diagnosis of the prediction of the behavior of the signals to generate the residual values, and applies to the construction of a decision tree from the most significant residue of a given signal, enabling the process of acquisition and formation of the signature matrix; **(ii) Ramoune et al.** [17] propose the modeling of the components of a gas turbine using a dynamic nonlinear autoregressive approach with external exogenous input NARX to monitor the vibrational dynamics of a turbine in operation. The results demonstrate the ability to detect and manage in real time possible failures caused mainly by intrinsic vibrations; **(iii) Liang et al.** [10], through real-time monitoring of killing operation parameters, propose the improved genetic optimization algorithm of a MLP ANN of the wellhead backpressure predictive control system by combining the prediction model with the pressure control model and use the engineer's method to study the bottomhole pressure control of a well; **(iv) Sanchéz et al.** [8] uses the parameter estimation technique for the detection and isolation of faults in the Development and Application of Methods for Actuator Diagnosis in Industrial Control Systems (DAMADICS) control valve, the mathematical model extracted for the valve was identified as a first-order Autoregressive with Extra Input (ARX) and a RBF. NN was added as complement to the conventional detector by identification of parameters, which allowed the correction of the error in the thresholds. They worked with 3 types of faults: blocking, sedimentation and erosion. For the 3 types of failures, the detector they propose obtained an efficiency of 1; **(v) Yun et al.** [22] propose a fault diagnosis method based on deep networks for diesel generator sets. Sensor data from the diesel generator set is collected and stored. They used a Deep Belief Network (DBN) to diagnose diesel engine failures, with a 7-layer network structure (1 input layer, 1 output layer and 5 hidden layers). They also compared their algorithm to others (MLP, ANN, Support Vector Machine (SVM) and Extreme Learning Machine(ELM)), finding that DBN is better to their scenario; **(vi) Kim and Heo** [9] used Linear Discriminant Analysis (LDA), Linear Regression (LR), SVM, DT, Random Forest (RF), XGBoost, LightGBM and MLP to detect normal and abnormal conditions for each hydraulic component and **(vii) Pang et al.** [15] propose a data-driven model and intelligent decision methodology for product quality control systems using Digital Twins concept and simulation methods. They incorporated MLP NN to ensure stability of the manufacturing process and improve the quality level of industrial valves.

Secondly, a summary of the selected works that propose methods based on ANN for supervised learning tasks is presented, **being LSTM one of the highlights**: **(i) Nie et al.** [13] propose the use of a Fuzzy Neural Network (FNN) and an integration between the AutoRegressive Integrated Moving Average (ARIMA) and the LSTM, called ARIMA-LSTM model, to predict the state and RUL of the water hydraulic High-Speed On/Off Valve (HSV), obtaining 93.30% accuracy to predict the state and 1.6% of relative error for predict the RUL; **(ii) Zhao et al.** [1] propose a Principal Component Analysis and Informer (PCA-Informer) model for the detection and prediction of nuclear valve operation failures to solve the problem of monitoring long time series, where PCA is applied to extract failure characteristics from sound and vibration signals. They also calculated some statistics, and then adapted a LSTM model based on PCA-Informer to predict the probable failure of the variable to which you should pay attention. PCA brought good results, but the deep learning-based model performed even better. **(iii) Zhao et al.** [23] propose a fault diagnosis method based on a hybrid neural network model composed of a Multi-scale CNN and LSTM (MCNNLSTM) algorithm for the main pump of the valve cooling system in a converter station and thus classify 4 types of failures (imbalance, looseness, parallel misalignment, angular misalignment) and a normal state. A precision of 0.987 was obtained in the test set, demonstrating that his method has good generalization capacity; **(iv) Wang et al.** [19] presents the implementation and optimization of deep learning-based RUL prediction models for electric gate valves in nuclear energy systems by combining a Convolutional Autoencoder (CAE) and a LSTM. The deeper features extracted by CAE and the original features are combined, which combination enriches the feature dimension and improves the predictive ability of LSTM; and **(v) Chompu and Kijsirikul** [4] propose the Attention-Based Bidirectional Long Short-Term Memory (ABD-LSTM) model using feature extraction along with data reduction techniques to predict failures in the WC, where they used PCA to reduce the dimensionality of the data and Light Gradient Boosting Machine (LightGBM) for the selection process of characteristics. They reduce the estimated memory usage needed to train LSTM-based models by 77.99 %. Their model achieved an optimal F1 score of 85.28 %. They also suggest applying Bidirectional LSTM (BDLSTM) type networks to classify time series. Thirdly, a summary of the selected works is presented, which also proposes methods based on ANN for supervised learning tasks, **being CNN one of the highlights**: **(i) Zhou et al.** [25] propose a One-Dimensional CNN (G-1DCNN) fault diagnosis model based on Gabor filter, which is used to improve the feature extraction ability of 1DCNN and improve the operating speed of the model without affecting the accuracy of fault diagnosis. The average diagnosis accuracy of the 1DCNN fault diagnosis model is as high as 99.26 % and 99.52 % respectively under different operating conditions (50 % and 100 %) of the actuator, and the detection accuracy of each fault reaches more than 98.5 %; **(ii) Wang et al.** [20] propose a two-story traceability identification method for high-voltage direct current (HVDC) system, based on wavelet entropy and Affinity Propagation(AP) clustering algorithm theory and CNN deep learning theory; **(iii) Mazaev et al.** [12] propose a Bayesian Convolutional Neural Networks (BCNN) based methodology for direct prediction of RUL of solenoid valves using current signals as training data; and **(iv) Griffiths** [6] propose an automated condition monitoring tool that uses self-encoding CNNs and t-distributed Stochastic Neighbor Embedding (t-SNE) for dimensionality reduction and intelligent feature selection, followed by Density-Based Spatial Clustering of Applications with Noise (DBSCAN) to identify variable groups of data.

Also, the selected works that proposes **methods based on other ML-based methods than ANNs for supervised learning tasks** are presented: **(i) Liu et al.** [11] propose a model for early diagnosis of reciprocating compressor valve failures based on Multiclass SVM and a DT. The results show that their model is better than models based on Neural Networks (NN) in the case of small samples, and its efficiency is greater than conventional SVMs; **(ii) Yugapriya et al.** [21] propose a method to determine the status of the hydraulic system based on sensor data (temperature, vibration, flame and voltage sensors connected through a Raspberry pi microcontroller). They used diverse ML algorithms: LR, K-Nearest Neighbors (KNN), DT, RF and Naive Bayes (NB). LR obtained the best accuracy with 93 %; **(iii) Pawar et al.** [16] uses fuzzy logic control and predictive analysis to determine control valve failures through vibration analysis, which can identify two types of failures and the percentage of failure occurrence while predictive analysis using RF gives the score based on the valve condition and identifies all types of failure events; **(iv) Bykor et al.** [3] propose a classifier model for predicting operating conditions of valves of a hydraulic test bench from an aircraft. They used 3 ML algorithms: eXtreme Gradient Boosting (XGBoost), KNN and SVM, which were trained with both standard parameters and hyperparameter correction through cross-validation sampling. The best result was the model based on the SVM algorithm with optimized hyper-parameters; and **(v) Van et al.** [18] propose three methods to detect faults in control valves: using the linear characteristic of the valve; using the valve reference model; and using Bagging algorithm based on DT. Their results show that the average fault detection accuracy of the linear feature is the highest (75.45 %), but it is not possible to classify specific faults and on the other hand the ML algorithm not only achieved an accuracy of (74.20 %), but also classifies specific faults.

Finally, Zhong et al. [24] propose a **method based on semi-supervised learning (SSL)**, the classification model is trained from a small amount of data with fault labels, thus generating pseudo labels for a large amount of unlabeled data for diagnosing hydraulic directional valve failures in operation. The contribution of such work is that a multi-sensor fusion algorithm is designed to obtain pseudo-labels with high confidence, and an adaptive threshold model similar to the generative countermeasure network is designed to intelligently generate thresholds to select pseudo-labels instead of intervention human, after 5 iterations, the average diagnosis reaches 99.72 % to 99.00 % accuracy.

4.1   Datasets

The different data sources used by the selected papers were analyzed. This is an important task as it is the first step for replicability. Only 5 (out of 23) used available datasets in the internet. Yugapriya et al. [21] uses a public dataset available in ZeMA gGmbH[1] This dataset is obtained from different sensors installed inside the test rig (PS1-PS6, ESP1, FS1- FS2, TS1-TS4, VS1, CE, CP, SE) condition variables (cooler condition, valve condition, internal pump leakage, hydraulic accumulator/bar, stable flag). Wang et al. [19] uses its own dataset and also a public one called NASA Turbofan Jet Engine Data Set available in B2[2], which the goal is to predict the RUL of each engine in the test. Bykor et al. [3] and Kim and Heo [9] use the same public dataset called Condition Monitoring of Hydraulic Systems available in Helwig et al.[3], which was obtained experimentally from a hydraulic test bench. Sanchéz et al. [8] uses the DAMADICS control valve model available in ValveActuator-DAMADICS[4]. The other works selected used datasets provided by private companies. Therefore, data source is not available in 18 out of 23 works. In summary, only 4 datasets are publicly available as 2 works used the same dataset.

5 Conclusions, Limitations and Future Work

The aim of this work was to understand and present the state of the art on the identification of valve anomalies using ML methods, as well as identify studies that aims to predict valves RUL. To this end, a SLR was conducted. As a result, a summary of the types of ML algorithms used by the selected studies is presented. It was observed that most of them aim to identify the RUL in valves by analyzing time series generated by different types of sensors contained in valves, and thus contribute to the preventive maintenance of industrial systems. It is also observed how different types of ML algorithms were used, from the simpler logistic regression to models constructed by deep learning algorithms, including LSTM and CNN. Hybrid ML models are also proposed to create more robust models. From this entire study, it was found that the most used are deep learning algorithms, among which ANN-based Methods stand out, followed by CNN-based Methods and LSTM-based Methods.

One limitation of this work is that It was not possible to compare the results among the analyzed works. This was due to (i) only 5 (out of 23) works publicly made available the datasets they used; (ii) the explored problems were modeled inherently different among them; (iii) the datasets were different; and (iv) the methods were developed for different tasks using different evaluation measures. However, it was observed that LSTM-based methods presented good prediction rates in some outstanding works, which may be due to inherent characteristics of time series, such as auto-correlation and dependency. However, much more should be done by the academic community in this way, as may be quite challenging for data scientists being aware of at least both time-series concepts, theories and techniques and ML and DL algorithms, concepts and theories.

---

[1] Available at https://universe.roboflow.com/zema-ggmbh/hrc-2.

[2] Available at https://www.kaggle.com/datasets/behrad3d/nasa-cmaps.

[3] Available at https://doi.org/10.24432/C5CW21.

[4] Available at https://github.com/dlaredo/ValveActuator-DAMADICS-.

As future lines of research, the following is presented: (i) Investing more research on long time prediction of RUL in valves, as it is a quite important task, specially into the Digital Twin scenario. For instance, it is observed that Zhao et al. [1] points out the importance of analysing long time series, which was not explored in the other works; (ii) It is also observed in the analyzed studies that there are a huge diversity of possible data pre-processing techniques for using ML algorithms and its importance to obtain better results. So, evolving frameworks for supporting Data Analysts work on anomalies prediction in valves could be quite helpful; (iii) As only 5 (of 23) works that were analyzed used datasets available on the Internet or made their data set available, it is understood that, for matter of reproducibility in order to evolve the knowledge in this theme, it is necessary to invest into collecting, treating and turning available more datasets related to industrial valves anomalies. This is specially important when considering that anomalies in industrial valves may mean high financial, environmental and social risks; (iv) Research could be conducted that focuses on the practical implementation of machine learning models in real industrial environments by fostering collaboration between academia, industry experts and valve system manufacturers. These collaborations can provide insights into practical challenges facing industrial environments and facilitate the development of more applicable and effective solutions.

References

1. An, Z., et al.: A novel principal component analysis-informer model for fault prediction of nuclear valves. Machines **10**(4), 240 (2022)
2. Andrade, A., Lopes, K., Lima, B., Maitelli, A.: Development of a methodology using artificial neural network in the detection and diagnosis of faults for pneumatic control valves. Sensors **21**(3), 853 (2021)
3. Bykov, A.D., Voronov, V.I., Voronova, L.I.: Machine learning methods applying for hydraulic system states classification. In: 2019 Systems of Signals Generating and Processing in the Field of on Board Communications, pp. 1–4 (2019)
4. Chomphu, W., Kijsirikul, B.: Wellhead compressor failure prediction using attention-based bidirectional LSTMs with data reduction techniques. In: Proceedings of the 2020 4th International Conference on Compute and Data Analysis, p. 16-22. ICCDA 2020 (2020)
5. Ghobakhloo, M.: Industry 4.0, digitization, and opportunities for sustainability. J. Cleaner Prod. **252**, 119869 (2020)
6. Griffiths, I.: Automated condition monitoring using artificial intelligence, ICVISP 2020 (2021)
7. Hu, W., Lim, K.Y.H., Cai, Y.: Digital twin and industry 4.0 enablers in building and construction: a survey. Buildings **12**(11), 2004 (2022)
8. José, S.A., Samuel, B.G., Arístides, R.B., Guillermo, R.V.: Improvements in failure detection of DAMADICS control valve using neural networks. In: 2017 IEEE Second Ecuador Technical Chapters Meeting (ETCM), pp. 1–5 (2017)
9. Kim, D., Heo, T.Y.: Anomaly detection with feature extraction based on machine learning using hydraulic system IoT sensor data. Sensors **22**(7), 2479 (2022)
10. Liang, H., Wei, Q., Lu, D., Li, Z.: Application of GA-BP neural network algorithm in killing well control system. Neural Comput. Appl. **33**, 949–960 (2021)

11. Liu, J., Yu, Z., Zhang, B., Hu, G., Chen, Z.: Early fault diagnosis model design of reciprocating compressor valve based on multiclass support vector machine and decision tree. Sci. Program. **2022**, 7486271 (2022)
12. Mazaev, G., Crevecoeur, G., Hoecke, S.V.: Bayesian convolutional neural networks for remaining useful life prognostics of solenoid valves with uncertainty estimations. IEEE Trans. Industr. Inf. **17**(12), 8418–8428 (2021)
13. Nie, S., Liu, Q., Ji, H., Hong, R., Nie, S.: Integration of ARIMA and LSTM models for remaining useful life prediction of a water hydraulic high-speed on/off valve. Appl. Sci. **12**(16), 8071 (2022)
14. Page, M.J., et al.: Prisma 2020 explanation and elaboration: updated guidance and exemplars for reporting systematic reviews. BMJ **372** (2021)
15. Pang, J., Zhang, N., Xiao, Q., Qi, F., Xue, X.: A new intelligent and data-driven product quality control system of industrial valve manufacturing process in CPS. Comput. Commun. **175**, 25–34 (2021)
16. Pawar, K.S., Sondkar, S., Dattarajan, S., Fernandes, N.: Comparative analysis of fuzzy logic and machine learning algorithm for predictive analysis of control valve. In: 2019 5th International Conference On Computing, Communication, Control And Automation (ICCUBEA), pp. 1–6 (2019)
17. Rahmoune, M.B., Hafaifa, A., Kouzou, A., Chen, X., Chaibet, A.: Gas turbine monitoring using neural network dynamic nonlinear autoregressive with external exogenous input modelling. Math. Comput. Simul. **179**, 23–47 (2021)
18. Van, K.T., Huynh, T.H., Dai, T.T., Nguyen, H.D., Vo, T.Q.: Real-time fault detection algorithms for industrial process control valve. In: 2022 6th International Conference On Computing, Communication, Control And Automation (ICCUBEA). pp. 1–7 (2022)
19. Wang, H., Jun Peng, M., Miao, Z., Kuo Liu, Y., Ayodeji, A., Hao, C.: Remaining useful life prediction techniques for electric valves based on convolution auto encoder and long short term memory. ISA Trans. **108**, 333–342 (2021)
20. Wang, Y., Tai, K., Song, Y., Kou, R., Zheng, Z., Zeng, Q.: Research on double-deck traceability identification method of commutation failure in HVDC system. IEEE Access **9**, 108392–108401 (2021)
21. Yugapriya, M., Judeson, A.K.J., Jayanthy, S.: Predictive maintenance of hydraulic system using machine learning algorithms. In: 2022 International Conference on Electronics and Renewable Systems (ICEARS), pp. 1208–1214 (2022)
22. Yun, Q., Zhang, C., Ma, T.: Fault diagnosis of diesel generator set based on deep believe network, pp. 186-190, AIPR 2019 (2019)
23. Zhao, Q., Cheng, G., Han, X., Liang, D., Wang, X.: Fault diagnosis of main pump in converter station based on deep neural network. Symmetry **13**(7), 1284 (2021)
24. Zhong, Q., et al.: Fault diagnosis of the hydraulic valve using a novel semi-supervised learning method based on multi-sensor information fusion. Mech. Syst. Signal Process. **189**, 110093 (2023)
25. Zhou, Z., Sun, J., Gao, W., Kang, J., Zhang, W.: Fault diagnosis of gas turbine actuator based on improved convolutional neural network. In: 2021 36th Youth Academic Annual Conference of Chinese Association of Automation (YAC), pp. 655–659 (2021)

Framework for Real-Time Predictive Maintenance Supported by Big Data Technologies

Marco Teixeira(✉), Francisco Thierstein, Pedro Entringer, Hugo Sá, José Demétrio Leitão, and Fátima Leal🔾

REMIT, Universidade Portucalense, Rua Dr. António Bernardino de Almeida, 4200-072 Porto, Portugal
{39952,41333,48123,38731,41479}@alunos.upt.pt, fatimal@upt.pt

Abstract. Industry 4.0 boosted the generation of large volumes of sensor data in manufacturing production lines. When adequately mined, this information can anticipate failures and launch maintenance actions increasing quality and productivity. This paper explores the integration of real-time big data techniques in industry. Specifically, this work contributes with a framework for real-time predictive maintenance supported by big data technologies. The proposed framework is composed of: (*i*) Apache Kafka as messaging system to manage sensor data; (*ii*) Spark as Machine Learning engine for large-scale data processing; and (*iii*) Cassandra as NoSQL distributed database. We showcase the synergy of these cutting-edge technologies in a predictive maintenance system tailored for the request. By leveraging advanced data analysis methods, we reveal hidden patterns and insights valuable for researchers across various disciplines. The experiments were performed with the NASA turbofan jet engine dataset, which includes run-to-failure simulated data from turbo fan jet engines.

Keywords: Big Data · Apache Kafka · Apache Spark · Cassandra · Real-time processing · Predictive Maintenance

1   Introduction

In today's data-driven world, the need for innovative and scalable solutions to handle and process large volumes of information is more critical than ever. This article aims to provide an in-depth exploration of big data, predictive maintenance, and the cutting-edge technologies that are transforming how we analyse and manage vast datasets.

Big data refers to the collection, processing, and analysis of massive and heterogeneous amounts of data [9]. One practical application of big data is predictive maintenance, which relies on data and Machine Learning (ML) algorithms to proactively anticipate equipment failures and schedule maintenance. This approach reduces downtime and costs while enhancing safety and efficiency. In addition, the exponential increase in the availability of streaming and time series

Á. Rocha et al. (Eds.): WorldCIST 2024, LNNS 985, pp. 13–22, 2024.
https://doi.org/10.1007/978-3-031-60215-3_2

data, due to the use of the Internet of Things (IoT) connected with real-time databases, is supporting a large number of applications with sensors that accept important and constantly changing data [1]. Developing a real-time anomaly detection system is crucial to ensure the early detection of failures in industrial equipment. The occurrence of equipment failures can lead to production stoppages, financial losses and even a risk to worker safety [15]. Early detection also ensures safety and efficiency in the industry, meeting maintenance costs and increasing equipment availability [1]. Therefore, the processing of large amounts of data streams has been explored in multiple domains of industry employing different disruptive technologies [9].

To address this challenge, *i.e.*, process large amounts of information as streams, we explore three key technologies that underpin modern big data solutions: (*i*) Apache Cassandra as NoSQL database; (*ii*) Spark as ML engine for large scale data processing; and (*iii*) Kafka as messaging system to manage multiple sensor data. Apache Cassandra is a highly-scalable, distributed NoSQL database designed for handling large quantities of structured and semi-structured data across multiple nodes. Apache Spark is an open-source, distributed computing system that excels in processing large datasets quickly and flexibly. Apache Kafka is a distributed streaming platform that enables real-time data processing.

The integration of these technologies creates a powerful data processing pipeline. Therefore, this paper proposes a framework for real-time predictive maintenance encompassing main 3 modules: (*i*) Kafka ingests and transmits data in real-time, feeding it into Spark; (*ii*) Spark for manipulation, analysis, and prediction using advanced algorithms; and (iii) Cassandra to store and manage the results, providing a scalable and reliable data storage solution. The experiments were performed with NASA's turbofan jet engines dataset predicting the Remaining Useful Life (RUL). The results show the suitability of this interconnected architecture for predictive maintenance and other time-sensitive applications. This article is organised as follows. Section 2 provides the literature review. Section 3 describes the proposed method. Section 4 reports the experimental results. Finally, Sect. 5 summarises and discusses the outcomes.

2   Related Work

An anomaly is a point in the dataset that is not associated with the normal behaviour of a system. Anomalies can be considered as rare events or observations that deviate significantly from the conventional behaviour or patterns observed in a single data point [4]. Over the past few years, computer science has made significant strides in all aspects related to data, encompassing data processing, transformation, and management. ML methods, which categorise, classify, and predict reality, have also seen rapid growth in the field of data analysis [7]. It is the case of predictive maintenance which has garnered significant attention within multidisciplinary research groups, propelling the creation and integration of research lines pertaining to data acquisition, infrastructure, storage, distribution, security, and intelligence. This multifaceted approach has been pivotal in comprehending and enhancing predictive maintenance systems, offering a holistic approach towards proactive fault management across various industrial sectors [17].

The impact of maintenance costs on the industry is significant, representing between 15% and 60% of total operating costs [6]. However, many companies do not adequately measure maintenance-related expenses, which justifies the need for studies that explore new technologies capable of changing this scenario. The implementation of predictive maintenance has been identified as a differentiator in the Industry 4.0 era, allowing the use of data collected by multiple sensors in industrial environments to predict the RUL of assets and, thus, reduce maintenance costs, minimise downtime, and improve productivity and quality of processes. Health management is one of the most important topics in the industry to predict the state of assets to avoid downtime and failures, as understanding when equipment will fail before it actually fails is valuable in several industries. In commercial aviation, predicting failures in important parts of the plane, e.g., the engines, can prevent disasters, save lives, and mitigate problems such as rescheduling several pilots and flight attendants due to changes in flight plans caused by conditions and aircraft failures. However, normally, it involves the treatment of large amounts of data in real-time, becoming a challenge.

The field of big data analytics has gained much attention in recent years due to the vast amounts of data generated by various domains, such as healthcare and weather forecasting. Big data analytics involves the processing and analysis of large volumes of data to derive insights, make predictions, and identify patterns. The use of technologies, e.g., Apache Kafka, Apache Spark, and Cassandra, can significantly enhance the efficiency and scalability of big data processing [13]. According to O'Donovan et al. [14], there is a significant increase in studies related to the integration of Big Data and Industry 4.0. Specifically, the research highlighted pertinent questions about the types of analyses employed and the areas of focus, especially in relation to maintenance and diagnostics, directly contributing to the approach of this study. The literature contemplates some solutions using big data technologies together with predictive models:

- Kanavos et al. [10] propose a healthcare platform that utilises reactive programming to provide real-time, predictive, and prescriptive analytics. The platform integrates Apache Kafka, Apache Spark, and Cassandra to process and analyse healthcare data in real-time. The platform's predictive modelling algorithms enable healthcare professionals to identify potential health risks in patients, while the prescriptive recommendations assist in making informed decisions for patient care.
- Ed-daoudy et al. [5] propose a scalable and real-time system for disease prediction. The system leverages Apache Kafka, Apache Spark, and Cassandra to collect, process, and analyse large volumes of healthcare data. The ML algorithms employed by the system enable it to predict the likelihood of diseases and provide real-time alerts to healthcare professionals.
- Kaur et al. [11] propose a big data framework for winter precipitation forecasting. The framework employs Apache Kafka, Apache Spark, and Cassan-

dra to process and analyse large volumes of weather data in real-time. The system's regularisation-based approach enables accurate predictions of winter precipitation and provides real-time alerts to weather forecasters.

- Nair *et al.* [12] present a health monitoring application that utilises Apache Spark and ML techniques to forecast users' health status based on data collected from Twitter. The prominent use of Apache Spark as a big data processing engine, coupled with the integration of Spark Streaming and MLib libraries, filters and analyses data in near real-time, enabling health status prediction.
- Asgari *et al.* [2] explore the utilisation of Apache Spark together with Hadoop to predict and map urban air quality. The study reveals that Spark, while operating on Hadoop, accesses data stored in the Hadoop Distributed File System (HDFS) and employs the YARN resource manager to distribute processing. This system was designed with a prediction module based on Spark, capable of training ML algorithms using historical air pollution monitoring data and subsequently forecasting air quality at monitoring stations.
- Canizo et al. [3] presents a Big Data approach for a predictive maintenance in wind turbines. The proposed method integrates Apache Spark, Apache Kafka, Apache Mesos, and HDFS providing a fault-tolerant scalable cloud computing environment solution.

Table 1 depicts a comparison of relevant related work that addresses predictive problems with a big data framework. The literature presents several approaches which employs a big data framework. However, they do not explore a stream-based big data approach for predictive maintenance. To overcome this gap, this work proposes a framework real-time predictive maintenance which employs Apache Cassandra as NoSQL database, Spark as ML engine for large scale data processing in real-time, and Kafka as messaging system to manage multiple sensor data.

Table 1. Comparison of contributor classification approaches.

Approach	Predictive Maintenance	Data Streams	Big Data Framework
Kanavos *et al.* [10]		✓	✓
Ed-daoudy *et al.* [5]		✓	✓
Kaur *et al.* [11]		✓	✓
Nair *et al.* [12]		✓	✓
Asgari *et al.* [2]		✓	✓
Canizo et al. [3]	✓	✓	✓
Our proposal	✓	✓	✓

3   Proposed Method

Real-time predictive maintenance in the industry requires advanced technologies and data analytics to predict equipment failures and schedule maintenance activities before issues occur. This paper proposes a real-time framework, illustrated in Fig. 1, for predictive maintenance to minimise downtime, reduce maintenance costs, and extend the lifespan of their assets. The framework includes 3 modules: (*i*) Apache Cassandra as NoSQL database for storage and data persistence; (*ii*) Spark as ML engine for large scale data processing; and (*iii*) Kafka as messaging system to manage multiple sensor data.

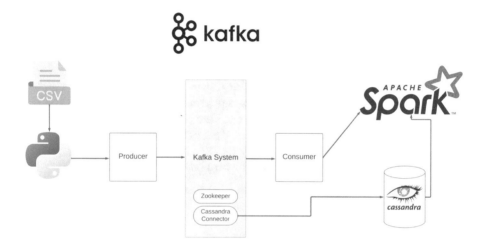

Fig. 1. Proposed framework for real-time predictive maintenance.

Storage and data persistence is done with Apache Cassandra. Due to Cassandra's capacity for tunability, it raises as a logical solution for systems which requires a durable and performant data store. Therefore, our framework combines Apache Cassandra and Apache Kafka to build a scalable, fault-tolerant, and real-time data processing. While Cassandra acts as a database to persist the processed data, the Kafka-Cassandra connector facilitates the transfer of messages from the Kafka topic to the Cassandra database. This approach ensures that the data remains securely stored and readily accessible for subsequent analysis.

Messaging system in the context of data handling, serves as a communication to exchange information among different components within a distributed network. In the context of a message system handling data from sensors, the fundamental objective is to establish a seamless and efficient connection between sensors and the broader data processing infrastructure. In this paper, we employ Apache Kafka to receive and manage data in real-time. Apache Kafka is a distributed streaming platform that is commonly used for building real-time data pipelines and streaming applications. It utilizes a publish-subscribe model, allowing producers to publish messages to topics, and consumers to subscribe to those topics and process the messages. The proposed framework, while the Kafka producer was utilized to send sensor data in the form of messages via a Kafka topic, the Kafka Client is used to receive the data and fit the ML model.

RUL Prediction is used to predict motor bearing failures. It relies on Spark framework for data streams. The data sensors information controlled and stored by Apache Kafka and Cassandra is retrieved to fed Spark. Due to the high quality of predictions proved by the literature review, we employ the Gradient Boosted Tree Regressor (GBTRegressor), Random Forest (RF), Decision Trees (DT), and Linear Regression to experiment the proposed framework.

Evaluation Protocol involves predictive accuracy metrics to measure the error between the predicted and the real values. It is the case of the Mean Absolute Error (MAE), which measures the average absolute deviation among the predicted rating and the real rating, or the Root Mean Square Error (RMSE), which highlights the largest errors [8]. The proposed method is empirically evaluated by calculating incrementally for each incoming event of sensors from Cassandra using an incremental RMSE proposed by Takács *et al.* (2009) [16].

4   Experiments and Results

We conducted several experiments with the NASA turbofan jet engine dataset[1] dataset to evaluate the proposed method. The data processing was implemented in Python using the Spark ML library[2]. Our system is running on a cloud OpenStack instance, holding 8 GB RAM, 8 CPU and 160 GB hard-disk.

4.1   Dataset

The experiments were performed with sensor data from a public NASA engine database. This dataset consists of four different sets covering two types of failures: (i) in the compressor; and (ii) in the compressor-propeller. In addition, it contemplates two general conditions: at sea level and in a different environment.

Each dataset is composed of a multivariate time series, divided into training and testing subsets.

In addition, three operational configurations are recorded holding a significant impact on engine performance. The data contains noise from the sensors, and in each time series, an engine operates normally initially, developing a failure

---

[1] Available at https://www.kaggle.com/datasets/behrad3d/nasa-cmaps, November 2023.

[2] Available at https://spark.apache.org/docs/latest/ml-classification-regression.html, November 2023.

at some point. In the training set, the failure magnitude increases progressively until system failure, while in the test set, the series ends before the point of complete failure. This framework provides important insights into the evolution of failures over time, providing a detailed view of engine behaviour.

The dataset encompasses 26 attributes categorizing the engines' operational cycles: Unit Number (Unique identifier for each engine unit), time in Cycles (Number of cycles elapsed since the start of measurement), operational settings 1, 2, and 3 (three different operational settings applied to the engines) and sensor measurements 1 to 23 (readings from 23 sensors capturing specific engine characteristics during operation). The dataset also provides the RUL values for the test data for performance evaluation.

4.2   RUL Prediction

RUL prediction was performed using GBTRegressor, RF, DT, and LR. We analysed the performance using MAE, RMSE, and R^2 to provide a quantitative assessment of the models. The framework contributes to the early detection of anomalies in engines in a big data scenario not compromising the efficiency of the model presented by the related work.

The performance evaluation within separated subsets FD001 to FD004, as illustrated in Table 2, delineates the prowess of the GBTRegressor model in predictive analytics. Across all subsets, GBTRegressor consistently displays superior predictive accuracy and explanatory power, as denoted by its lowest MAE of 21.136, RMSE of 29.396, and highest R^2 at 0.811 in FD001, followed by competitive metrics in subsequent subsets. Despite its commendable performance, GBTRegressor requires higher training times, notably 7.661 in FD001 and reaching 9.780 in FD004. These findings highlight the model's robustness in capturing variance within the data, yet emphasise the critical trade-off between enhanced predictive performance and increased computational demands, crucial considerations when selecting a model for deployment in varied applications.

Table 3 presents the performance metrics combining the multiple subsets. Among the assessed algorithms, the GBTRegressor notably exhibits superior predictive efficacy, reflected by its minimal MAE of 37.695 and RMSE of 51.439, surpassing other models. Furthermore, the GBTRegressor demonstrates a commendable R^2 at 0.669, denoting its enhanced capacity in explaining data variability. However, it is crucial to note the considerable variance in training durations, with RF and DT presenting lower training times at 2.985 and 3.626, respectively, in contrast to the longer training time of GBTRegressor at 10.579. Despite these variations, all models display comparable prediction times. Consequently, the selection of an ideal model depending of subset requires a comprehensive consideration of not only performance metrics but also broader contextual factors, encompassing model scalability, interpretability, and alignment with specific problem requisites.

Table 2. Regression Models Performance Metrics by subset

Dataset	Algorithms	MAE	RMSE	R^2	Training Time	Prediction Time
FD001	DT	25.771	36.0201	0.717	4.401	0.156
	GBTRegressor	**21.136**	**29.396**	**0.811**	7.661	0.078
	LR	30.373	39.349	0.662	2.267	0.736
	RF	25.720	35.659	0.723	**2.116**	**0.062**
FD002	DT	26.503	36.620	0.723	4.21	0.078
	GBTRegressor	**21.341**	**29.442**	**0.821**	7.036	0.109
	LR	30.687	39.727	0.674	2.142	0.641
	RF	25.829	36.098	0.731	**1.844**	**0.062**
FD003	DT	38.358	54.646	0.696	4.422	0.093
	GBTRegressor	**26.596**	**36.973**	**0.861**	8.847	0.080
	LR	30.587	39.789	0.623	2.369	0.685
	RF	36.551	51.932	0.723	**2.239**	**0.078**
FD004	DT	45.189	60.546	0.557	4.940	0.093
	GBTRegressor	**38.437**	**52.157**	**0.671**	9.780	0.093
	LR	42.972	56.418	0.615	**2.238**	0.812
	RF	43.568	58.003	0.593	2.723	**0.078**

Table 3. Regression Models Performance Metrics combing the multiple subsets.

Algorithms	MAE	RMSE	R^2	Training Time	Prediction Time
DT	44.902	60.032	0.550	3.626	0.109
GBTRegressor	**37.695**	**51.439**	**0.669**	10.579	0.109
LR	42.700	55.992	0.608	5.673	1.046
RF	43.046	57.118	0.592	**2.985**	**0.078**

5 Conclusions

Predictive maintenance harnesses advanced analytics to anticipate equipment failures and schedule maintenance proactively, aiming to minimize downtime and reduce costs. One of the significant challenges in implementing predictive maintenance lies in managing and analyzing vast volumes of diverse data generated by sensors and machinery in real time, emphasizing the importance of effective big data strategies for actionable insights. In this context, this paper contributes to a framework for real-time predictive maintenance supported by big data technologies. The framework is composed of 3 modules: (*i*) Apache Cassandra as NoSQL database for storage and data persistence; (*ii*) Spark as ML engine for large-scale data processing; and (*iii*) Kafka as a messaging system to manage multiple sensor data.

The ability to process data in real-time and apply advanced analytics techniques enabled accurate predictions of critical failures in rocket motor bearings. The data storage approach implemented using Cassandra also yielded impressive results. The database provides a reliable and efficient means of persisting the processed data, ensuring its availability for subsequent analysis. Furthermore, the architecture employed throughout the project demonstrates scalability. It seamlessly handled large volumes of data, showcasing its robustness and ability to support high data throughput. Apache Kafka helps to manage the sensor information as data streams. Finally, Spark provides the ML library to employ the required model for RUL prediction.

For projects with a smaller amount of data, this architecture may prove to be overly burdensome in terms of maintenance costs and computational resources. The infrastructure required to maintain and operate such a system could outweigh the benefits for smaller-scale projects. On the other hand, for projects involving a substantial amount of data, this architecture demonstrated robustness and redundancy. Its ability to handle large-scale data processing and provide a high level of data integrity is perfect for ensuring the prevention of data loss.

To sum up, this paper demonstrates the feasibility of receiving sensor data in a streaming fashion and applying an ML model to predict critical failures in rocket motor bearings. The combination of big data technologies, streaming data processing, and advanced analytics showcased the potential for proactive maintenance and enhanced operational efficiency in the aerospace industry. As future work, we intend to explore: (i) different configurations in spark architecture using multiple nodes for predictions; (ii) test with different datasets; (iii) automatic detection of environment selecting the proper ML model, accordingly.

Acknowledgements. This work was supported by the UIDB/05105/2020 Program Contract, funded by national funds through the FCT I.P.

References

1. Ahmad, S., Purdy, S.: Real-time anomaly detection for streaming analytics. arXiv preprint arXiv:1607.02480 (2016)
2. Asgari, M., Farnaghi, M., Ghaemi, Z.: Predictive mapping of urban air pollution using apache spark on a hadoop cluster. In: Proceedings of the 2017 International Conference on Cloud and Big Data Computing, pp. 89–93 (2017)
3. Canizo, M., Onieva, E., Conde, A., Charramendieta, S., Trujillo, S.: Real-time predictive maintenance for wind turbines using big data frameworks. In: 2017 IEEE International Conference on Prognostics and Health Management (ICPHM), pp. 70–77. IEEE (2017)
4. Chatterjee, A., Ahmed, B.S.: IoT anomaly detection methods and applications: a survey. Internet Things **19**, 100568 (2022). https://www.sciencedirect.com/science/article/pii/S2542660522000622
5. Ed-daoudy, A., Maalmi, K., El Ouaazizi, A.: A scalable and real-time system for disease prediction using big data processing. Multimedia Tools Appl. **82**, 30405–30434 (2023)

6. Haarman, M., Mulders, M., Vassiliadis, C.: Predictive maintenance 4.0: predict the unpredictable. PwC and Mainnovation 4 (2017)
7. Han, J., Pei, J., Tong, H.: Data Mining: Concepts and Techniques. Morgan Kaufmann, Waltham (2022)
8. Herlocker, J.L., Konstan, J.A., Borchers, A., Riedl, J.: An algorithmic framework for performing collaborative filtering. In: Proceedings of the 22nd Annual International ACM SIGIR conference on Research and Development in Information Retrieval, pp. 230–237. ACM (1999)
9. Jin, X., Wah, B.W., Cheng, X., Wang, Y.: Significance and challenges of big data research. Big Data Res. 2(2), 59–64 (2015). https://www.sciencedirect.com/science/article/pii/S2214579615000076, visions on Big Data
10. Kanavos, A., Trigka, M., Dritsas, E., Vonitsanos, G., Mylonas, P.: A regularization-based big data framework for winter precipitation forecasting on streaming data. Electronics 10(16), 1872 (2021)
11. Kaur, J., Mann, K.S.: AI based healthcare platform for real time, predictive and prescriptive analytics using reactive programming. In: 10th International Conference on Computer and Electrical Engineering, Univ Alberta, Edmonton, CANADA, 11–13 October 2017 (2018). Journal of Physics Conference Series, vol. 933
12. Nair, L.R., Shetty, S.D., Shetty, S.D.: Applying spark based machine learning model on streaming big data for health status prediction. Comput. Electr. Eng. 65, 393–399 (2018)
13. Oussous, A., Benjelloun, F.Z., Ait Lahcen, A., Belfkih, S.: Big data technologies: a survey. J. King Saud Univ. Comput. Inf. Sci. 30(4), 431–448 (2018). https://www.sciencedirect.com/science/article/pii/S1319157817300034
14. O'donovan, P., Leahy, K., Bruton, K., O'Sullivan, D.T.: Big data in manufacturing: a systematic mapping study. J. Big Data 2, 1–22 (2015)
15. Shaqrah, A., Almars, A.: Examining the internet of educational things adoption using an extended unified theory of acceptance and use of technology. Internet Things 19, 100558 (2022)
16. Takács, G., Pilászy, I., Németh, B., Tikk, D.: Scalable collaborative filtering approaches for large recommender systems. J. Mach. Learn. Res. 10, 623–656 (2009)
17. Zonta, T., da Costa, C.A., da Rosa Righi, R., de Lima, M.J., da Trindade, E.S., Li, G.P.: Predictive maintenance in the Industry 4.0: a systematic literature review. Comput. Ind. Eng. 150, 106889 (2020). https://www.sciencedirect.com/science/article/pii/S0360835220305787

Emotional Evaluation of Open-Ended Responses with Transformer Models

Alejandro Pajón-Sanmartín[1], Francisco de Arriba-Pérez[1] (ID),
Silvia García-Méndez[1] (ID), Juan C. Burguillo[1] (ID), Fátima Leal[2] (ID),
and Benedita Malheiro[3,4]([envelope]) (ID)

[1] Information Technologies Group, atlanTTic, University of Vigo,
Campus Universitario de Vigo, Lagoas-Marcosende, 36310 Vigo, Spain
{apajon,farriba,sgarcia}@gti.uvigo.es, J.C.Burguillo@uvigo.es
[2] REMIT, Universidade Portucalense, 4200-072 Porto, Portugal
fatimal@upt.pt
[3] ISEP, Polytechnic of Porto, Rua Dr. António Bernardino de Almeida, 431,
4249-015 Porto, Portugal
mbm@isep.ipp.pt
[4] INESC TEC, Campus da Faculdade de Engenharia da Universidade do Porto,
4200-465 Porto, Portugal

Abstract. This work applies Natural Language Processing (NLP) techniques, specifically transformer models, for the emotional evaluation of open-ended responses. Today's powerful advances in transformer architecture, such as ChatGPT, make it possible to capture complex emotional patterns in language. The proposed transformer-based system identifies the emotional features of various texts. The research employs an innovative approach, using prompt engineering and existing context, to enhance the emotional expressiveness of the model. It also investigates spaCy's capabilities for linguistic analysis and the synergy between transformer models and this technology. The results show a significant improvement in emotional detection compared to traditional methods and tools, highlighting the potential of transformer models in this domain. The method can be implemented in various areas, such as emotional research or mental health monitoring, creating a much richer and complete user profile.

Keywords: Emotional analysis · GPT-3.5 · spaCy · Transformers · Prompt engineering · Language-based Emotion Recognition

1 Introduction

Natural Language Processing (NLP) models have gained significant importance in recent years, primarily due to advances in their ability to analyse and understand human language in an automated manner [6]. These improvements have been propelled and the technology popularised by companies such as OpenAI[1] and Google[2].

---

[1] Available at https://openai.com, reviewed in January 2024.
[2] Available at https://ai.google, reviewed in January 2024.

© The Author(s), under exclusive license to Springer Nature Switzerland AG 2024
Á. Rocha et al. (Eds.): WorldCIST 2024, LNNS 985, pp. 23–32, 2024.
https://doi.org/10.1007/978-3-031-60215-3_3

This research uses NLP techniques for emotional evaluation of open-ended responses. To achieve this goal, transformer models with power to understand the human psyche are employed [7], becoming valuable tools in a wide range of fields such as psychology [8], education [12], stock market [17], or marketing [4].

The proposed method explores the ChatGPT model (GPT-3.5) [2] together with prompt engineering (as discussed in Sect. 4.1) to control response generation. These techniques are still experimental, as there are no rules specifying an exact input/output relationship, only recommendations to consider when using a model. Tests are conducted using spaCy for polarity detection, concluding with a comparison between both models. The aim is to provide an effective and straightforward solution for emotion detection to be deployed in any domain with minimal modification, something that does not currently exist.

This paper is structured as follows. Section 2 delves into the state of the art in natural language processing, Sect. 3 focuses on the objectives, and Sect. 4 details the methodology for analysing conversations using ChatGPT and spaCy. Section 5 briefly explores the applications of this approach for analysing extensive texts, highlighting the differences between the models. Finally, Sect. 6 presents the conclusions and outlines future steps.

2   Related Work

NLP has grown significantly in recent years, primarily due to its flexibility to adapt to wide range of applications and to generate complex responses with minimal instructions. This research explores the ability of NLP models to perform emotional evaluation, aiming to comprehend and analyse the emotions expressed in a text. The most prominent techniques are as follows.

Linguistic feature extraction, used by initiatives such as the Semantic Orientation Calculator [14], assigns polarities to different words, creating a dictionary, and applies several algorithms to calculate emotional scores for each entry, resulting in a final classification.

Supervised machine learning employs classification algorithms, such as Naive Bayes [19], Support Vector Machines [20], or Neural Networks [18], to train models with large labelled data sets to classify texts into emotional categories.

The most advanced and sophisticated solutions are based on transformers, which capture representations of words and contexts in a general and flexible way, adding significant value. Prominent models in emotional detection include Bidirectional Encoder Representation Transformers (BERT) [1] and Generative Pre-trained Transformers (GPT) [9], among others. Most of these solutions are proprietary, such as Anthropic[3], the basis for the enterprise conversational assistant Claude[4], or Inflection[5], a model used to create a personal intelligence assis-

---

[3] Available at https://www.anthropic.com/, reviewed in January 2024.
[4] Available at https://www.anthropic.com/index/introducing-claude, reviewed in January 2024.
[5] Available at https://inflection.ai/about, reviewed in January 2024.

tant. There are also open source approaches, such as Vicuna[6], a modified version of the Large Language Model Meta AI (LLaMA) [15] from Meta.

Given the high flexibility of transformer-based models, they have been applied in various domains, including virtual assistants (BingChat or ChatGPT), machine translation [13], automatic text summarising [5], semantic search [10], and emotional analysis [11]. However, most of these solutions are code-oriented, which significantly restricts their reuse by researchers or professionals. The primary advantages of the current proposal over existing methods lie in its simplicity and seamless adaptability to diverse environments. Furthermore, it harnesses advanced and diverse models, ensuring a high level of precision.

3   Objectives

The main objective of this project is to develop a robust system capable of analysing emotions and polarities in any text or human interaction with the greatest possible accuracy. Specifically, it aims to:

- **Detect emotions and polarities with GPT-3.5:** The GPT-3.5 NLP model will be used to: (i) detect and classify emotions and polarities; and (ii) identify the most prominent *topics* in each interaction for user profiling.
- **Detect polarities with spaCy:** The spaCy NLP library will be used to detect and analyse existing polarities.
- **Compare spaCy and GPT-3.5 results:** Comparisons will be made regarding performance, *i.e.* the ability to correctly capture text polarity.

4   Proposed Method

The designed method explores prompt engineering with the help of GPT-3.5 and spaCy in two contexts: interactive conversations and extensive texts. The considered emotions – Joy, Anger, Aversion, Sadness, Surprise, Fear, and Neutral – are based on the primary emotions model of Ekman and Cordaro [16], whereas polarity includes the Negative, Positive, and Neutral labels.

4.1   Prompt Engineering

Prompt engineering, a technique currently under development [3], is used to control the results of generative artificial Intelligence (AI) models. A prompt is then a natural language textual description of the instructions or keywords given to guide a generative AI model. As a central tool in natural language processing, it can be used in a wide variety of applications.

---

[6] Available at https://lmsys.org/blog/2023-03-30-vicuna/, reviewed in January 2024.

4.2   Use of GPT-3.5

In a memory based approach, conversations are used as input data. The entire conversation is submitted to the model, indicating in the prompt the input to analyse in each iteration. This decision to provide the maximum possible context allows the model to obtain deeper results since past events are important [21]. To create a modular system, the iterative prompt creation process divides the text into eight fragments. Since the goal is to successfully detect emotions, polarities and keywords, the model is explicitly provided with the objective and parameters of the analysis (Fig. 1a).

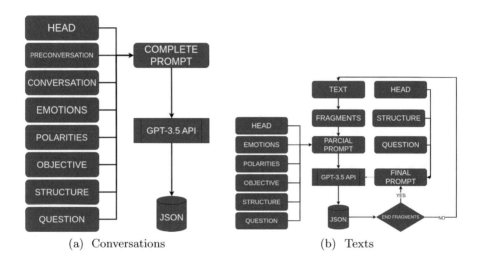

(a) Conversations   (b) Texts

Fig. 1. Diagram of the analysis process with GPT-3.5

The analysis of a conversation, according to Fig. 1a, includes the following steps: (i) assembly of all the fragments of the prompt; (ii) addition of the complete conversation; and (iii) specification of the conversation interaction to be analysed by the model (question). Then, there are other prompt fragments used to control the model and obtain the desired output, *e.g.*, emotions, polarities, objectivity, and structure. This process is repeated for each interaction[7] belonging to the conversation. The data are submitted via OpenAI's Application Programming Interface (API). Requests are made recursively until a satisfactory response is obtained. The results obtained by the model are processed and stored in a JSON[8] file, containing the polarity, emotion, and topics of each interaction.

The analysis of extensive texts, *e.g.*, from interviews or transcriptions, is summarised in Fig. 1b. In many cases, due to the limitation of 4096 *tokens*[9] per

---

[7] A conversation comprises one or more interspersed interactions between speakers.

[8] Available at https://www.json.org/json-en.html, reviewed in January 2024.

[9] Word used to designate a set of input characters submitted to the model.

model of GPT-3.5, it is impossible to segment texts based on paragraphs. Therefore, the text is dynamically divided into fewer fragments than this maximum. The prompt is also adapted, while maintaining the same modular philosophy. The process of analysing each fragment is identical to that of the conversations, sending each fragment sequentially. However, due to the *tokens* limitation, it is impossible to add the full text along with the fragment at this stage. In the last step, the JSON file holding the analysis of all the fragments is sent with a specific prompt that allows the data to be extracted from the full text. As a result, the final JSON file holds the complete analysis of the text. A dynamic adjustment of the request size has been performed. In the first call, the maximum size of 4096 *tokens* is maintained; however, after five requests, it automatically adjusts to the average between the total length of the prompt and the response.

4.3   Use of SpaCy

The polarity analysis was performed with spaCy[10], a Python library for NLP. Specifically, it employs a light and popular polarity detection pipeline for English texts – the `spaCyTextBlob`[11] – based on the `TextBlob` library. The adopted pipeline loads the medium-sized English model trained on written web text. Therefore, all entries must be translated into English.

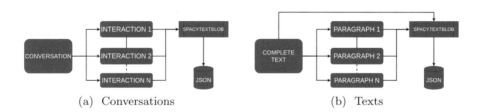

(a) Conversations   (b) Texts

Fig. 2. Diagram of the analysis process with spaCy

The mode of operation is very similar to the one adopted with GPT-3.5. First, spaCy processes the interactions of a conversation. In this case, the process is simplified by not having to create the prompt, since the only input is the text, as can be seen in Fig. 2a. Next, the model generates an output between -1 and 1, which is interpreted as positive, negative, or neutral if it is 0. Finally, the output is stored in a JSON file. In the case of the longer and more complex TED talk texts, they were translated for greater precision beforehand using the `textBlob`[12] translation API. Then, spaCy processes the entire text and the paragraphs independently since there has no entry size limitation.

---

[10] Available at https://spacy.io, reviewed in January 2024.
[11] Available at https://spacy.io/universe/project/spacy-textblob reviewed in January 2024.
[12] Available at https://textblob.readthedocs.io/en/dev/, reviewed in January 2024.

5   Experiments and Results

The results were evaluated by calculating the weighted average of the *Precision*, *Accuracy*, *Recall* and *F1-score* metrics. To facilitate the interpretation of the results, the respective confusion matrices are also presented.

Since spaCy only detects polarity, the emotions and polarity analysis were performed independently with the Conversations[13], TED talks[14], and the Short phrases[15] data sets. Table 1 details the contents of these data sets. Prior any experiments, the Conversations and TED talks data sets were manually labelled by the authors, who have an NLP background.

Table 1. Contents of the data sets

Data set	Labels	Number of Entries	Words per Entry		
			Average	Minimum	Maximum
Conversations	No	1494	14	1	153
TED talks	No	2801	17	1	199
Short phrases	Yes[a]	14448	9	3	34

[a] Includes only Positive and Negative polarity labels.

5.1   Emotion Analysis

The emotion analysis was performed with the three data sets, using GPT-3.5[16]. The Conversations data set comprises an average of 14 words per interaction. As can be seen in Fig. 3a, the results are positive since no attempt has been made to identify a contrary emotion, achieving an average *F1-score* of 71 %. The metrics improve with the simplest emotions, such as Joy or Fear, and fall moderately in more complex ones, such as Surprise.

The Spanish TED talks transcriptions contain an average of 17 words per paragraph. The results (Fig. 3b) are more stable than the previous ones. Since the fragments are longer, on average 1735 words per talk, they provide more context, improving the emotional analysis to an average *F1-score* of 84 %.

Finally, the method was applied to the short phrases data set with an average of 9 words per sentence. The complexity of this analysis was higher since these are

---

[13] Available at https://www.kaggle.com/datasets/projjal1/human-conversation-training-data, reviewed in January 2024.

[14] Available at https://www.kaggle.com/datasets/miguelcorraljr/ted-ultimate-dataset, reviewed in January 2024.

[15] Available at https://huggingface.co/datasets/hita/social-behavior-emotions, reviewed in January 2024.

[16] The model hyperparameters were set to `temperature=0.0`, `top_p=1.0` (default value), `frequency_penalty=0.0` (default value), `presence_penalty=0.0` (default value) and `stop_sequence=None` (default value).

phrases that have an inherent feeling, which results in a very complex labelling. Figure 3c shows that the results are positive and the wrong classifications are easily explainable. In complex emotions such as Surprise, confusion occurs with Joy, something that is not strange given the close relationship of these feelings. These disturbances cause performance to drop to an average *F1-score* of 62 %.

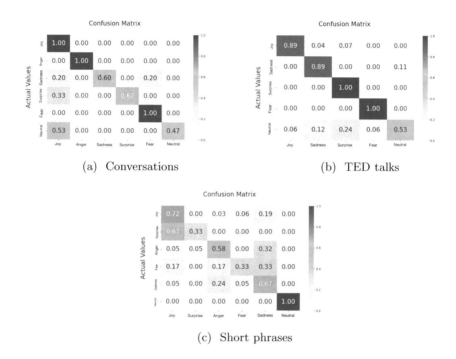

(a) Conversations

(b) TED talks

(c) Short phrases

Fig. 3. Emotional analysis confusion matrices

These results, summarised in Table 2, show an *F1-score* performance above 70 % with two of the data sets. The majority of the wrongly labelled emotions correspond to related feelings, *i.e.* few were classified as antonyms.

Table 2. Emotion analysis with GPT-3.5

Data set	Precision	Recall	F1-score
Conversations	0.85	0.72	0.71
TED talks	0.86	0.85	0.84
Short phrases	0.62	0.61	0.62

5.2   Polarity Analysis

Polarities have also been identified on all data sets with GPT-3.5 and spaCy (Tables 3 and 4). In the case of conversations, both models present good results, with GPT-3.5 obtaining an average *F1-score* value of 78 % and spaCy 66 %. The performance of spaCy with the TED talks drops to an average *F1-score* of 40 %, while that of GPT-3.5 increases to 88 % (Fig. 4). The larger and more complex fragments affect negatively spaCy, because ambiguity typically increases as the sentence size increases, and favour GPT-3.5, due to richer context.

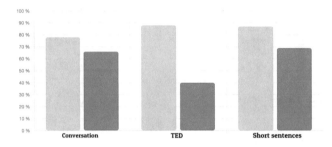

Fig. 4. Polarity analysis results (*F1-score*) with GPT-3.5 (green) and spaCy (rose)

The Short phrases results deepen the differences between both models. GPT-3.5 scores an average *F1-score* of 87 %, while spaCy scores 69 %. However, neutral polarities were ignored because they were not labelled in this data set.

These results indicate that both GPT-3.5 and spaCy can be used for polarity detection. However, there is a clear distinction in their recommended scope, depending on the length of the input. spaCy can be employed with smaller text fragments and GPT-3.5 with longer ones.

Table 3. Polarity analysis with GPT-3.5

Data set	Precision	Recall	F1-score
Conversations	0.89	0.77	0.78
TED talks	0.89	0.89	0.88
Short phrases	0.85	0.88	0.87

Table 4. Polarity analysis with spaCy

Data set	Precision	Recall	F1-score
Conversations	0.66	0.67	0.66
TED talks	0.49	0.39	0.40
Short phrases	0.69	0.69	0.69

6   Conclusions

This work explores the use of NLP techniques for emotional evaluation on open-ended responses. To this end, it uses transformer models together with prompt engineering to control the generation of responses. The emotion (GPT-3.5) and polarity results (GPT-3.5 and spaCy) were obtained with three data sets.

GPT-3.5 displayed better performance when compared with spaCy, specially with long texts. Transformer models produce a very precise analysis thanks to the identification of semantic relationships and the size of their databases. The use of prompt engineering was more complex than expected, since a large number of tests with different instructions had to be resorted to obtain an output that met the expected results. It was possible to obtain very different results with practically identical instructions, leading to errors in the subsequent processing. These problems were solved with fine tuning, such as limiting the number of *tokens* or increasing the value of the `top_p` hyperparameter used by GPT-3.5 to control the diversity of the generated text.

While the `spaCyTextBlob` pipeline is limited to polarity detection, the GPT-3.5 model performs a complete analysis of emotions, polarities and topics, which is essential for comprehensive applications. This proposal stands out for its advantages, particularly its adaptability and ease of use for non-technical users.

The future expansion of this emotion detection groundwork offers opportunities to create detailed psychological profiles and improve the ability to recognise self-destructive behaviour and identify early signs of neurological disorders through the integration of specialised algorithms. It is also necessary to further study the application of this method to other use cases, as each model has different limitations.

Acknowledgement. This work was partially supported by: (*i*) Xunta de Galicia grants ED481B-2021-118 and ED481B-2022-093, Spain; and (*ii*) Portuguese national funds through FCT – Fundação para a Ciência e a Tecnologia (Portuguese Foundation for Science and Technology) – as part of project UIDB/50014/2020 (DOI: 10.54499/UIDP/50014/2020 | https://doi.org/10.54499/UIDP/50014/2020).

References

1. Al-Omari, H., Abdullah, M.A., Shaikh, S.: EmoDet2: emotion detection in English textual dialogue using BERT and BILSTM models. In: 2020 11th International Conference on Information and Communication Systems (ICICS), pp. 226–232. IEEE (2020)
2. Brown, T., et al.: Language models are few-shot learners. In: Advances in Neural Information Processing Systems. vol. 33, pp. 1877–1901. Curran Associates, Inc. (2020)
3. Dale, R.: GPT-3: what's it good for? Nat. Lang. Eng. **27**(1), 113–118 (2021)
4. Hartmann, J., Netzer, O.: Natural language processing in marketing. In: Artificial Intelligence in Marketing, vol. 20, pp. 191–215. Emerald Publishing Limited (2023)
5. Kai, W., Lingyu, Z.: Research on text summary generation based on bidirectional encoder representation from transformers. In: 2020 2nd International Conference on Information Technology and Computer Application (ITCA), pp. 317–321 (2020)
6. Khurana, D., Koli, A., Khatter, K., Singh, S.: Natural language processing: state of the art, current trends and challenges. Multimedia Tools Appl. **82**(3), 3713–3744 (2023)

7. Liu, Y., et al.: A survey of visual transformers. IEEE Trans. Neural Netw. Learn. Syst., 1–21 (2023)
8. Mann, P., Matsushima, E.H., Paes, A.: Detecting depression from social media data as a multiple-instance learning task. In: 2022 10th International Conference on Affective Computing and Intelligent Interaction (ACII), pp. 1–8 (2022)
9. Mao, R., Liu, Q., He, K., Li, W., Cambria, E.: The biases of pre-trained language models: an empirical study on prompt-based sentiment analysis and emotion detection. IEEE Trans. Affect. Comput. **14**(3), 1743–1753 (2023)
10. Masuda, K., Matsuzaki, T., Tsujii, J.: Semantic search based on the online integration of NLP techniques. Procedia Soc. Behav. Sci. **27**, 281–290 (2011). Computational Linguistics and Related Fields
11. Peng, S.: A survey on deep learning for textual emotion analysis in social networks. Digit. Commun. Netw. **8**(5), 745–762 (2022)
12. Shen, J.T., et al.: MathBERT: a pre-trained language model for general NLP tasks in mathematics education (2023)
13. Sun, Z., Wang, M., Li, L.: Multilingual Translation via Grafting Pre-trained Language Models (2021)
14. Taboada, M., Brooke, J., Tofiloski, M., Voll, K., Stede, M.: Lexicon-based methods for sentiment analysis. Comput. Linguist. **37**(2), 267–307 (2011)
15. Touvron, H., et al.: LLaMA: open and efficient foundation language models (2023)
16. Tracy, J.L., Randles, D.: Four models of basic emotions: a review of Ekman and Cordaro, Izard, Levenson, and Panksepp and Watt. Emot. Rev. **3**(4), 397–405 (2011)
17. Wang, C., Chen, Y., Zhang, S., Zhang, Q.: Stock market index prediction using deep transformer model. Expert Syst. Appl. **208**, 118128 (2022)
18. Wehrmann, J., Becker, W., Cagnini, H.E.L., Barros, R.C.: A character-based convolutional neural network for language-agnostic twitter sentiment analysis. In: 2017 International Joint Conference on Neural Networks (IJCNN), pp. 2384–2391 (2017)
19. Wongkar, M., Angdresey, A.: Sentiment analysis using Naive Bayes algorithm of the data crawler: Twitter. In: 2019 Fourth International Conference on Informatics and Computing (ICIC), pp. 1–5 (2019)
20. Zainuddin, N., Selamat, A.: Sentiment analysis using Support Vector Machine. In: 2014 International Conference on Computer, Communications, and Control Technology (I4CT), pp. 333–337 (2014)
21. Zhou, Y., Kang, X., Ren, F.: Prompt consistency for multi-label textual emotion detection. IEEE Trans. Affect. Comput. **15**, 1–10 (2023)

Big Data Analytics in the Supply Chain: A Bibliometric Review of Scholarly Research from 2012 to 2023

Samuel Fosso-Wamba[1]([⊠]) (ID), Cameron Guthrie[1] (ID),
and Maciel Manoel de Queiroz[2] (ID)

[1] TBS Business School, Toulouse, France
s.fosso-wamba@tbs-education.fr
[2] FGV EAESP, Sao Paulo, Brazil

Abstract. The professional and scholarly interest in big data analytics in the supply chain has rapidly grown over the past decade. This article presents a bibliometric analysis of academic literature to improve our understanding of the evolution and current state of the field. The analysis covers 920 published journal and conference papers identified using the Web of Science database. A descriptive analysis of the corpus first allows us to identify the most influential works, ideas, and authors in the field, leading to the identification of opportunities for future research.

Keywords: Big data analytics · supply chain · bibliometric review

1 Introduction

Big data analytics (BDA) is an important enabling tool for supply chain (SC) optimization. Indeed, the BDA-enabled SC has been shown to lead to important benefits, including improved prediction of future customer demand, enhanced end-to-end SC efficiency, and visibility, reduced SC costs, improved decision-making, and ultimately improved SC performance and competitive advantage. All these BDA-enabled SC benefits drive the BDA market size in the SC. Today, the BDA-enabled SC market is expected to grow from $3.66 Billion in 2020 to about $13.37 Billion in 2028 [1].

Early studies on BDA-enabled SC have already demonstrated its capacity to transform SCs towards higher business value creation and capture. For example, Chen et al. [2], using data collected from SC managers from 157 North American organizations, found that BDA use for SC optimization was directly associated with better decision-making capability. Fosso Wamba and Akter [3] also found that BDA-enabled SC significantly improved firm performance. Hung et al. [4] found that in the Asian banking sector, BDA-enabled SC improved SC finance and the efficiency of marketing tactics and campaigns. Jaouadi [5] argued that BDA capability could predict SC innovation. In this study, we seek to describe the intellectual structure of the BDA-enabled SC extant literature by answering the following research questions:

Á. Rocha et al. (Eds.): WorldCIST 2024, LNNS 985, pp. 33–41, 2024.
https://doi.org/10.1007/978-3-031-60215-3_4

- What is the present level of research on BDA-enabled SC?
- Which authors are leading the field in production and citations?
- Which journals and countries are leading the field?
- What are potential future directions of BDA-enabled SC research?

The paper is organized as follows. The first section presents and defines the key concepts. The methodology used to collect and analyze the literature is then outlined. The current state of the literature is then described followed by an analysis of research patterns. The paper concludes with a discussion of possible avenues for future research and the limitations of the study.

2   Big Data Analytics in the Supply Chain

BDA applications have emerged as a disruptive paradigm in the field of SC. They have been shown to help build and support competitive advantage [6, 7], mainly through their predictive power [8] and other analytical techniques [9]. Thus, BDA can improve firm [3, 10] and supply chain performance [11, 12].

On the one hand, traditional approaches to store and explore data (e.g., data mining or data warehousing) are no longer sufficient. On the other hand, with advances in the digital transformation of organizations [7], supply chains are challenged to develop and effectively manage robust strategies to add business value [13] by exploiting the large quantities of data that are being continuously generated in the SC. Consequently, BDA has become an essential tool [7] to analyze the massive quantities of data produced by organizations both within and outside of the supply chain. Not surprisingly, BDA has garnered high interest from a wide range of practitioners, scholars, and policymakers.

3   Methodology

In this study, we use bibliometric techniques to describe the current state of the BDA for SC extant literature. We set up inclusion and exclusion criteria so that the most relevant articles were extracted from the Web of Science database. As this is a young field, all journal articles, conferences, and editorials related to big data analytics in the supply chain were included. A total of 920 documents were retrieved using the search terms "big data analytics" AND "supply chain"[1].

The Bibliomerix R package was used to calculate descriptive statistics and bibliometric measures such as citation counts [14] and VOSviewer software was used for network visualizations [15].

4   Results

In this section we use the results of the bibliometric analysis to describe the current state of accumulated research. Descriptive statistics of the collection are provided in Table 1.

---

[1] The search was conducted on November 24, 2023.

Table 1. Main information regarding the collection

Description	
Documents	920
Sources	260
Period	2012–2023
Annual growth rate	86%
Average citations per document	40.5
Authors	2322
Authors of single-authored documents	46
Documents per author	0.39
Co-authors per document	3.53
International co-authorships	51.6%

A total of 2322 authors contributed to the literature. The documents were mainly co-written with an average of 3.53 co-authors per document. The publications were spread between 2012 to 2023, with the earliest publication a paper by [16] in the Journal of Systems Science and Systems Engineering describing how the confluence of big data analytics, adaptive services and digital manufacturing enables mass customization.

Figure 1 reports the annual production of documents. We can see that the number of documents increased rapidly from 2018 onwards.

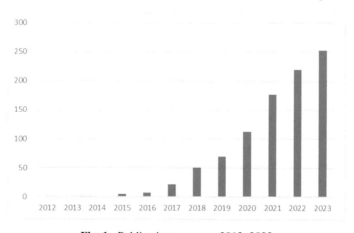

Fig. 1. Publications per year 2012–2023

Table 2 presents the five most influential publications by total number of citations in the BDA and supply chain. The table shows that the top-ranked paper is by Zhong R. Y. et al. (2017), entitled "Intelligent manufacturing in the context of industry 4.0: a review," with 1176 citations and 168 citations per year. The second most influential publication is by Oztemel E. & Gursev S. (2020), entitled "Literature review of Industry 4.0 and related technologies," with 754 citations and 188.5 citations per year. The third most influential publication is by Wang G. et al. (2016) on big data analytics in logistics and supply chain management, and Ivanov D. et al. (2019) is the fourth most influential publication, and the fifth most influential publication is by Hu, H., Wen, Y., Chua, T. S., & Li, X. (2014). The table reveals a variety of topics related to BDA in SC, including big data analytics, digital supply chains, and IoT applications, highlighting the significant academic interest and diverse research activity in this area.

Table 2. Most influential papers

Paper	Citations	Citations/year
1. Zhong, R. Y., Xu, X., Klotz, E., & Newman, S. T. (2017). Intelligent manufacturing in the context of industry 4.0: a review. *Engineering*, *3*(5), 616–630	1176	168
2. Oztemel, E., & Gursev, S. (2020). Literature review of Industry 4.0 and related technologies. *Journal of intelligent manufacturing*, *31*, 127–182	754	188.5
3. Wang, G., Gunasekaran, A., Ngai, E. W., & Papadopoulos, T. (2016). Big data analytics in logistics and supply chain management: Certain investigations for research and applications. *International journal of production economics*, *176*, 98–110	697	87.1
4. Ivanov, D., Dolgui, A., & Sokolov, B. (2019). The impact of digital technology and Industry 4.0 on the ripple effect and supply chain risk analytics. *International journal of production research*, *57*(3), 829–846	690	138
5. Hu, H., Wen, Y., Chua, T. S., & Li, X. (2014). Toward scalable systems for big data analytics: A technology tutorial. *IEEE access*, *2*, 652–687	577	57.7

Table 3 presents the most influential sources in BDA in SC research, described by their total number of publications and citations, their h-index, and the earliest year of publication on these topics. Following common practice in bibliometric studies, the sources are ranked by h-index. The h-index is a measure of both the quantity and the quality of scholarly output. It reports the total number of publications that have received the same number of citations. In case of a draw, sources are ranked by the total number of citations.

We can see from Table 3 that while the International Journal of Production Research has fewer publications than Annals of Operations Research (i.e., 37 vs. 54), it leads the table with its total contributions since 2017: 37 papers have been published, accumulating 4,369 citations. This has contributed to the journal's h-index of 27 for BDA in SC research. The second ranked source is Technological Forecasting and Social Change, with 34 publications since 2018, 2,444 citations, and a h-index of 23, followed by Annals of Operations Research, which has the highest number of publications in the table (i.e., 54 since 2018). The International Journal of Production Economics is ranked in fourth position, with 25 publications since 2015, 2775 citations, followed by Production Planning & Control in fifth position with 25 publications since 2017, 1136 citations and a h-index of 17.

Table 3. Most influential sources.

Source	P	C	h	Y
1. International Journal of Production Research	37	4369	27	2017
2. Technological Forecasting and Social Change	34	2444	23	2018
3. Annals of Operations Research	54	1375	20	2018
4. International Journal of Production Economics	25	2775	17	2015
5. Production Planning & Control	25	1136	17	2017

P = Publications, C = Citations, h = h-index, Y = Start year

Table 4 ranks the most influential authors by h-index. Once again, the total citations are used to rank authors in the case of a draw. The table shows that Gunasekaran A. from the USA leads the list with 25 publications, 3991 citations, and an h-index of 20, starting in 2016. Then, Fosso-Wamba S. from France is in the second position with 17 publications, 1739 citations, an h-index of 15. Bag S. from South Africa is in the third position with 20 publications, 1415 citations and an h-index of 15, starting in 2016. In fourth position is Dubey R from France with 14 publications, 1959 citations, and a h-index of 13. The list also features authors from India, the United Kingdom, and France, that have significantly contributed to the field from 2018 onwards.

Table 4. Most influential authors

Author	P	C	h	Co	Y
Gunasekaran A	25	3991	20	USA	2016
Fosso-Wamba S	17	1739	15	France	2016
Bag S	20	1415	15	South Africa	2019
Dubey R	14	1959	13	France	2016
Gupta S	23	1128	13	France	2019
Kamble S	16	1602	12	India	2018
Mangla SK	11	594	11	United Kingdom	2019
Foropon C	11	1166	10	France	2019
Dwivedi Y	14	906	10	United Kingdom	2020
Raut R	10	482	10	India	2019

P = Publications, C = Citations, h = h-index, Co = Country, Y = Start year

Table 5 presents an overview of publications by country in BDA and supply chain research. Countries are ranked by their h-index in the field. The data is based on the first author's country at the time of publication. From the table, we can observe that the United Kingdom is leading with 111 articles, a h-index of 40, and an MCP ratio of 0.48. China is ranked in second position with 164 publications, a h-index of 40, but fewer total citations than UK scholars. India follows in the third position with 113 articles and a relatively low MCP of 0.35. Australia and France are ranked in fourth and fifth positions respectively, with significantly lower productivity and influence in the field compared to the top ranked countries. However, both countries have higher MCP ratios indicating a strong inclination towards international collaborations. This is also the case for seventh ranked Brazil (MCP = 0.78). Finally, the USA, Italy, Germany, and Canada are also active in the field, albeit at lower levels of influence and productivity. This table highlights the global spread and collaborative nature of research in BDA and supply chain.

Table 5. Most productive countries

Country	P	C	h	Top 50	Y	MCP
United Kingdom	111	6968	40	6	2016	0.68
China	164	6336	40	8	2014	0.56
India	113	3728	32	3	2015	0.35
Australia	72	3596	29	3	2017	0.76
France	58	2853	26	5	2016	0.76
USA	37	1635	20	8	2012	0.03
Brazil	32	1441	19	0	2017	0.78
Italy	40	1499	17	4	2017	0.30
Germany	34	1546	16	5	2017	0.21
Canada	30	766	11	1	2017	0.70

P = Publications, C = Citations, h = h-index, Top 50 = Publications amongst 50 most cited, Y = Start year, MCP = Proportion of multiple country publications

We next examined the conceptual themes researched by scholars by grouping together co-occurring keywords. Figure 2 presents the resulting network map based on 180 keywords that were used at least 10 times in the dataset.

The size of each node represents keyword frequency while the lines represent the co-occurrence of keywords in publications. We observe seven thematic clusters, three of which account for over 80% of all keywords used in the corpus and merit examination.

The first (green) cluster accounts for 37% of all keyword occurrences, grouping together keywords relating to analytics, supply chain management, information, and big data. The second (red) cluster groups together keywords relating to predictive analytics, capabilities, competitive advantage, and firm performance. It accounts for 28% of keyword occurrences. The third (blue) cluster concerns sustainability, industry 4.0 and

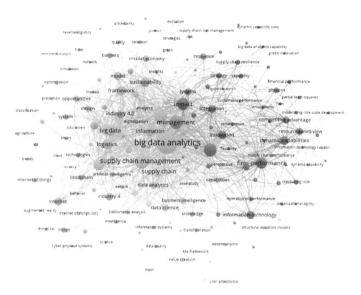

Fig. 2. Co-occurrence network of keywords

associated technologies (e.g., blockchain, artificial intelligence) and circular economy. It accounts for 17% of keyword occurrences in the corpus.

5   Discussion and Implications

We began this study by seeking to describe the intellectual structure of the literature on BDA-enabled SC. Our analysis provides some insights about the present level of research in the field. It shows that the annual growth rate has been 86%, and from 2012–2023 920 documents were published on the topic with the most part being a result of international collaboration.

Regarding the authors that are leading this field, we ranked the most influential authors using their h-index and total citations. The results show that Gunasekaran A. from the USA leads the field with 25 publications, and an h-index of 20, followed by Fosso-Wamba S. from France with 17 publications, 1739 citations, an h-index of 15. Then, Bag S. from South Africa, with 20 publications, 1415 citations and an h-index of 15, starting in 2016, is in the third position. In fourth position, Dubey R from France has 14 publications, 1959 citations, h-index of 13.

Regarding journals and countries that are leading the field of research in BDA-enabled SC, the International Journal of Production Research is the main outlet with publications since 2017. It's followed by Technological Forecasting and Social Change, and Annals of Operations Research, which had the highest number of publications during the period. International Journal of Production Economics is ranked fourth, followed by Production Planning & Control.

To rank countries, we use their h-index in BDA and supply chain research. The results based on the first author's country show that scholars in the UK are the most productive and influential, followed by China and India. Australia and France follow with high MCP ratios indicating strong international collaborations with scholars from these countries. Brazil has the highest MCP in the table. Finally, the USA, Italy, Germany, and Canada have also contributed to the field, albeit with less influence.

Our results provide important insights into the extant literature in BDA and supply chain. They allude to a set of future research directions. For example, our findings suggest that while BDA gained popularity in SC in manufacturing contexts, mainly anchored by the industry 4.0 paradigms, perspectives have since shifted to more holistic views, following digital transformation trends. For instance, topics related to BDA in support of resilience, sustainable performance, traceability, disaster management, innovation, capabilities, and servitization could be further studied by scholars. Future studies could also explore the joint adoption and use of BDA and artificial intelligence in the SC. The human and ethical dimensions in BDA-enabled SC adoption, integration and optimization is also an interesting avenue for future research.

6 Conclusions

Our research has provided an overview of the evolution and current state of the field of big data analytics in the supply chain. We used bibliometric and network analysis of 920 published journal and conference papers identified using the Web of Science database. Our analysis has allowed us to highlight high citation topics, collaborative patterns between institutions and countries, and possible avenues for future research into BDA-enabled SC.

References

1. VM Intelligence: Supply Chain Big Data Analytics Market Size and Forecast (2022). https://www.verifiedmarketresearch.com/product/global-supply-chain-big-data-analytics-market-size-and-forecast-to-2025/. Accessed 24 Nov 2023
2. Chen, D., Preston, D., Swink, M.: How big data analytics affects supply chain decision-making: an empirical analysis. J. Assoc. Inf. Syst. **22**, 1224–1244 (2021)
3. Fosso Wamba, S., Akter, S.: Understanding supply chain analytics capabilities and agility for data-rich environments. Int. J. Oper. Prod. Manage. **39**(6/7/8), 887–912 (2019)
4. Hung, J.-L., He, W., Shen, J.: Big data analytics for supply chain relationship in banking. Ind. Mark. Manage. **86**, 144–153 (2020)
5. Jaouadi, M.H.O.: Investigating the influence of big data analytics capabilities and human resource factors in achieving supply chain innovativeness. Comput. Ind. Eng. **168**, 108055 (2022)
6. Talwar, S., et al.: Big Data in operations and supply chain management: a systematic literature review and future research agenda. Int. J. Prod. Res. **59**(11), 3509–3534 (2021)
7. Queiroz, M.M., Fosso Wamba, S.: Managing the Digital Transformation: Aligning Technologies, Business Models, and Operations. CRC Press, Boca Raton (2023)
8. Phillips-Wren, G., Hoskisson, A.: An analytical journey towards big data. J. Decis. Syst. **24**(1), 87–102 (2015)
9. Gandomi, A., Haider, M.: Beyond the hype: big data concepts, methods, and analytics. Int. J. Inf. Manage. **35**(2), 137–144 (2015)

10. Fosso Wamba, S., et al.: The performance effects of big data analytics and supply chain ambidexterity: the moderating effect of environmental dynamism. Int. J. Prod. Econ. **222**, 107498 (2020)
11. Waller, M.A., Fawcett, S.E.: Data science, predictive analytics, and big data: a revolution that will transform supply chain design and management. J. Bus. Logist. **34**(2), 77–84 (2013)
12. Bansal, P., Gualandris, J., Kim, N.: Theorizing supply chains with qualitative big data and topic modeling. J. Supply Chain Manag. **56**(2), 7–18 (2020)
13. Schoenherr, T., Speier-Pero, C.: Data science, predictive analytics, and big data in supply chain management: current state and future potential. J. Bus. Logist. **36**(1), 120–132 (2015)
14. Aria, M., Cuccurullo, C.: Bibliometrix: an R-tool for comprehensive science mapping analysis. J. Informet. **11**(4), 959–975 (2017)
15. van Eck, N., Waltman, L.: Software survey: VOSviewer, a computer program for bibliometric mapping. Scientometrics **84**(2), 523–538 (2009)
16. Tien, J.M.: The next industrial revolution: integrated services and goods. J. Syst. Sci. Syst. Eng. **21**(3), 257–296 (2012)

Human-Machine Interface Based on Electromyographic (EMG) Signals Aimed at Limb Rehabilitation for Diabetic Patients

Hubet Cárdenas-Isla(✉), Bogart Yail Márquez(✉), Ashlee Robles-Gallego(✉), and José Sergio Magdaleno-Palencia(✉)

Instituto Tecnológico de Tijuana, Calzada Del Tecnológico S/N, Fraccionamiento Tomas Aquino, 22414 Tijuana, Baja California, México
{m23210003,bogart,ashlee.robles,jmagdaleno}@tectijuana.edu.mx

Abstract. Diabetes is a condition derived from high blood sugar levels for a prolonged period. Which triggers several complications such as anemia, blindness, erectile dysfunction, cardiovascular problems, high blood pressure, poor circulation, and a high probability of gangrene when there is a skin cut on an extremity where the most viable solution to risks is go septicemia or blood poisoning by pathogens generated by dead tissue is amputation of the limb. For such patients, where the limb amputation has already been carried out, tailored solutions are generated with the corresponding rehabilitation by learning to use this new tool to give them independence and inclusivity in their daily lives. Now, technology such as electromyographic sensors is required to read the electrical pulses generated by the muscles and then with artificial intelligence learn to interpret the electrical signals from the muscles of the patient's amputated limb. The main objective of this literature is to relate and obtain a prediction of the diabetes risk index of the general population, based on the incidence data of diabetic disease obtained by delegation in the United States, Mexico. -us through public health institutions. For the above, the prediction will be carried out by a multilayer perceptron neural network and an exploration of human-robot interface (HRI) solutions supported by electromyographic sensors will be carried out.

Keywords: Human-machine interface · human-robot interface (HRI) · electromiagraphy (EMG) · signal recognition · Diabetes

1 Introduction

Actually, we can find several electromyography solutions, so electromyography (EMG) reflects the electrical activity of muscle fibres during contraction, and it has been widely used for intelligent prostheses or exoskeleton robotics control [1, 7]. Electromyography (EMG) has gained prominence as a human-robot interface (HRI), particularly in the realm of collaborative wearable robots [9]. EMG, a noninvasive signal capturing the electrical activity of skeletal muscles without causing harm to the skin, faces susceptibility to disturbance during acquisition. Therefore, for an optimal EMG signal, it necessitates

Á. Rocha et al. (Eds.): WorldCIST 2024, LNNS 985, pp. 42–50, 2024.
https://doi.org/10.1007/978-3-031-60215-3_5

amplification and filtering using high-precision acquisition equipment [10]. However, the resulting high-precision EMG signal proves impractical for the direct control of wearable robots. This is because the control system must navigate numerous signal features, demanding hardware with a high signal sampling rate and memory to ensure the real-time performance of human-robot interaction (HRI) [11]. Human-computer interfaces are considered vital in the design of a limb exoskeleton prosthesis through which amputees can interact and control a specific machine [12].

So, decades of research have shown that diabetes affects part of the vulnerable, that is, low-income, adult population, with relatively intractable patterns observed in the increased risk of diabetes and rates of diabetes complications and mortality in these populations [13]. Diabetes has become one of the most serious and common chronic diseases of our time, causing costly, disabling, and life-threatening complications, as well as reducing life expectancy [14]. However, the impact of diabetes on bone health is frequently ignored or underestimated [19].

Finally, this material works with diagnoses of diabetic patients in Mexico, obtained from publicly accessible data sources and through a multilayer perceptron neural network to calculate the risk index of the patients.

2   State of Art

EMG signals have proven to be a promising modality for implementing human-machine (HHM) interfaces and controlling robotic devices, especially in the field of smart prostheses and limb rehabilitation (Liao et al. [1]). EMG reflects the electrical activity of motor units during muscle contraction, providing real-time information about movement intention that can be used to control robotic systems intuitively (Zhang et al. [7]). In patients who have suffered limb amputation because of severe complications of diabetes, EMG makes it possible to detect residual muscle activation patterns to control the movement of a robotic prosthesis that replaces the lost limb (Miften et al. [12]). In this way, the EMG interface offers these patients the possibility of restoring a certain degree of functionality and independence.

Implementing EMG-based HHM interfaces in robotic rehabilitation involves a series of signal processing to extract relevant features to infer user intent and identify specific muscle activation patterns. This requires the use of advanced algorithms and machine learning techniques to achieve robust and real-time decoding of the EMG signal (Liu et al. [11]). Recently, deep learning has been applied to EMG signal processing to improve pattern recognition, achieving more intuitive and natural control of robotic prostheses by users (Ferreira et al. [8]). Integrating inertial sensors and other perception systems also enriches contextual information for more advanced hybrid control.

Beyond prosthetic control, the EMG interface has also been explored as a modality for muscle rehabilitation in patients with motor sequelae of diabetes (Bai et al. [9]). Through the biofeedback provided by the EMG interface in real-time, the patient can modulate the activation of the affected muscle groups and gradually recover strength and functionality with the assistance of a robotic system. This EMG-guided rehabilitation approach allows personalized and adaptive therapies to be designed according to each patient's condition. Although significant progress has been made, some technological challenges remain for

EMG interfaces to reach their full potential in rehabilitation applications and improve the quality of life of patients with severe motor disabilities due to diabetes (Islam et al. [10]). The variability in EMG signals between users and conditions requires more robust and adaptive algorithms. Likewise, more compact, and portable designs for interfaces are needed. Effective integration with prosthetic systems and exoskeletons for smooth and safe control also represents an opportunity.

EMG interfaces have proven to be a promising technology for restoring functionality and independence in amputee patients, as well as for robotic rehabilitation in cases of motor disability due to diabetes. However, further research into advanced signal processing algorithms and systems integration is required to translate the potential of this interface into viable clinical solutions that improve patients' quality of life. As these challenges are overcome, EMG interfaces are positioned as an invaluable tool in the future of rehabilitation and medicine [25–31].

3   Justification of the Study

Development of electromyographic (EMG)-based human-machine interfaces for limb rehabilitation in diabetic patients, driven by the imperative to create technological solutions that enhance the quality of life and functionality of individuals who have undergone amputations due to severe diabetes complications. EMG interfaces detect movement intentions through residual muscle activation patterns, facilitating control over prosthetic devices replacing the lost limb. The objective is to restore a certain level of independence and enable the performance of daily activities for these patients. The necessity for advanced signal processing algorithms and system integration is emphasized to translate the potential of this technology into viable clinical solutions. Therefore, this work is justified by its focus on a relevant public health issue, where technology can significantly impact the rehabilitation and improvement of patients' quality of life.

This research provides a comprehensive overview of the current state of research regarding using electromyographic (EMG) interfaces for limb rehabilitation in the clinical context of diabetic patients. It delves into the intricacies of EMG signal processing for feature extraction and the recognition of relevant patterns. The discussion encompasses the application of advanced algorithms and machine learning for EMG-based prosthetic control and decoding. The article highlights the potential of EMG interfaces not only for prosthetic device control but also for muscular rehabilitation in cases of motor disability due to diabetes. It identifies current technological challenges in EMG signal processing and its effective integration with prosthetic systems and exoskeletons. However, the limitations of the article include its reliance on a literature review and the lack of experimental results on the use of EMG interfaces in diabetic patients. Furthermore, it doesn't delve into specific technical details of signal processing or machine learning algorithms for EMG, omitting performance assessments, usability, or clinical viability in humans. The article also neglects discussions on ethical, regulatory, or commercial implications related to the development and implementation of this technology, and it does not explore the multimodal integration of EMG with other interface modalities or perceptions for advanced hybrid control.

As shown in Fig. 1, data from the IMSS database shows that the recorded annual instances of detected diabetes cases in each Mexican state from 2000 to 2022 exhibit a

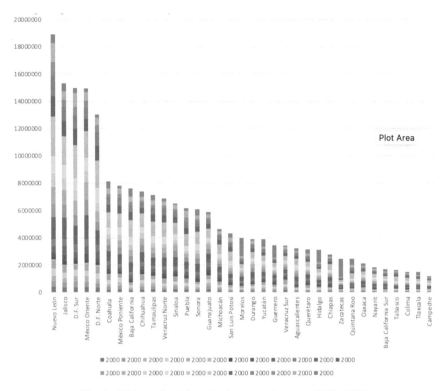

Fig. 1. Diabetes by delegation between the years 2000–2022.

general upward trajectory. The dataset covers all 32 states and Mexico City, segmented into three zones. Notably, states such as Nuevo León, Jalisco, Mexico City (across its three zones), Mexico State, Baja California, and Coahuila consistently report higher case numbers. However, a marked reduction in cases is observed in 2022, which could be attributed to the ramifications of the COVID-19 pandemic. The increasing trend in case detection over the years suggests a potential rise in disease prevalence and improvements in detection and diagnostic strategies. To provide a comprehensive perspective, exploring the incidence rate per 100,000 inhabitants is recommended, acknowledging the varying population sizes of each state. In conclusion, the data underscores the escalating challenge of diabetes as a public health concern in Mexico, emphasizing regional variations and hinting at the potential influence of external events, such as the ongoing COVID-19 pandemic.

4   Materials and Methods

As we know artificial intelligence has been used (AI) has been used widely in the health area to assist professionals into making the right decisions with a focus on screening classifications and diagnostics, among other possibilities [8].

Artificial intelligence (AI) is a fast-growing field and its applications to diabetes, can reform the approach to diagnosis and management of this chronic condition. Principles of machine learning have been used to build algorithms to support predictive models for the risk of developing diabetes or its consequent complications. AI allows continuous and burden-free remote monitoring of the patient's symptoms and biomarkers [16]. Further, social media and online communities enhance patient engagement in diabetes care. Technical advances have helped to optimize resource use in diabetes. Together, these intelligent technical reforms have produced better glycemic control with reductions in fasting and postprandial glucose levels, glucose excursions, and glycosylated hemoglobin [17]. AI will introduce a change in basic assumptions in diabetes care from conventional management strategies to building targeted data-driven precision care [15–17].

Digital medicine, digital research, and artificial intelligence (AI) have the power to transform the field of diabetes with continuous and no-burden remote monitoring of patients' symptoms, physiological data, behaviours, and social and environmental contexts through the use of wearables, sensors and smartphone technologies [18].

For use in this paper, we took free access data from health dependencies that are concentrated in each municipality. Our research does not involve human intervention or collection biological samples since we already use the data from databases provided by the National Institute of Public Health and other dependencies. The years we took the data for are from a period between 2000 through 2022. And we used the multilayer perceptron neural network [20–24] to analyze the data as we know artificial intelligence has been used (Fig. 2).

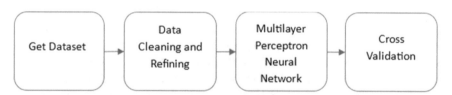

Fig. 2. Process model.

5 Results

The data we used are directly taken from the Mexican Institute of social Security (IMSS) [2, 3]. This information was meticulously treated and analyzed through a complex configuration that allows taking advantage of input, weights and intrinsic biases, obtaining the following results:

With an average of 70.3% training and 29.7% testing, a prediction average of 65.4% was obtained for the multilayer perceptron neural network. Consequently, we have an average high-risk level of 21.2%, an intermediate risk level of 17.3%, and a low-risk level of 61.5% (Fig. 3).

The mathematical representation of this model is given by Eq. 1:

$$Y = f(WxT + b) \tag{1}$$

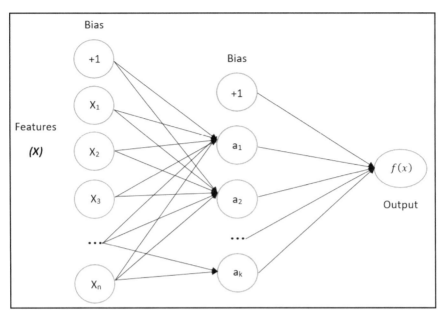

Fig. 3. Multilayer perceptron network with one hidden layer [24].

"Where W is the parameter, or weights, in the layer, x is the input vector, which can also be the output of the previous layer, and b is the bias vector" [5, 6]. In this case, we have 23 inputs, cover a period from 2000 to 2022 year. The output of the system makes known 3 levels: high-risk level, intermediate level, and finally low-risk level risk.

6   Conclusion

By observing the results obtained, we can conclude that the generated prediction shows a higher average of low risk. This leads us to focus on both high and low risk averages to offer them as a population segmentation to address solutions for limb rehabilitation in patients with diabetes in this case. Even if the government has already created some initiatives to fight the expansion of the problem posed by diabetes, our data shows that the prevalence and mortality of diabetes will continue to grow in the coming years.

Therefore, we must approach the development of tailored solutions for patients in a more effective and accessible way. How can it be to generate bionic gear prostheses with EMG sensors and artificial intelligence with their respective rehabilitation to learn to use this new tool, especially in patients who have undergone an amputation of their limbs. In addition, to the part of the population already affected by the disease, provide personalized assistance programs to guarantee a more pleasant and inclusive life.

Consequently, future work would be to develop a bionic gear prosthesis that is tailored to the patient, that is, a device that is functional, practical and safe or of medical grade. For the above, the type of use that will be given to it would have to be evaluated, whether rough or for a more modest use. Also, the lifestyle and conditions of the user such as his age, for example. Since, if it is a minor patient, updates to the device must be made as the patient grows.

Acknowledgement. A thank deeply to the authors and researchers who collaborated with their advice and recommendations in the preparation of the present material. Sincerely thank to the Instituto Tecnológico de Tijuana for their invaluable support and Instituto Mexicano del Seguro Social (IMSS) for their providing generalousy the necessary information to this research.

References

1. Liao, Z., et al.: Human–robot interface based on sEMG envelope signal for the collaborative wearable robot. Biomimetic Intell. Robot. (2023). https://doi.org/10.1016/j.birob.2022.100079
2. IMSS: Detección de Diabetes, por deledación (2023). https://datos.gob.mx/busca/dataset/deteccion-de-diabetes-por-delegacion
3. IMSS: istabla43_2022 - Detección padecimientos Diabetes por delegación, por año 2000 - 2022 (2023). http://datos.imss.gob.mx/dataset/informacion-en-salud/resource/60f146be-1528-4abe-8ecc-5daf8f8ca05c
4. Costa, L., et al.: Multilayer Perceptron. Introduction to Computational Intelligence, 105
5. Kumar Kain, N.: Understanding of Multilayer perceptron (MLP), 21 November 2018. https://medium.com/@AI_with_Kain/understanding-of-multilayer-perceptron-mlp-8f179c4a135f#:~:text=Each%20layer%20is%20represented%20as,b%20is%20the%20bias%20vector
6. Yao, S.-W., Ullah, N., Rehman, H.U., Hashemi, M.S., Mirzazadeh, M., Inc, M.: Dynamics on novel wave structures of non-linear Schrödinger equation via extended hyperbolic function method. Results Phys. **48**, 106448 (2023). https://doi.org/10.1016/j.rinp.2023.106448
7. Zhang, J., Zhao, Y., Shone, F., Li, Z., et al.: Physics-informed deep learning for musculoskeletal modeling: predicting muscle forces and joint kinematics from surface EMG. Neural Syst. (2022). https://ieeexplore.ieee.org/abstract/document/9970372/
8. Ferreira, A.C.B.H., et al.: Neural network-based method to stratify people at risk for developing diabetic foot: a support system for health professionals. PLoS ONE **18**(7), e0288466 (2023)
9. Bai, S., Islam, M.R., Power, V., OŚullivan, L.: User-centered development and performance assessment of a modular full-body exoskeleton (AXO-SUIT). Biomimetic Intell. Robot. **2**(2), 100032 (2022)
10. Islam, M.R.U., Waris, A., Kamavuako, E.N., Bai, S.: A comparative study of motion detection with FMG and sEMG methods for assistive applications. J. Rehabil. Assist. Technol. Eng. **7**, 2055668320938588 (2020)
11. Liu, J., Wang, C., He, B., Li, P., Wu, X.: Metric learning for robust gait phase recognition for a lower limb exoskeleton robot based on sEMG. IEEE Trans. Med. Robot. Bionics **4**(2), 472–479 (2022)
12. Miften, F.S., Diykh, M., Abdulla, S., Siuly, S., Green, J.H., Deo, R.C.: A new framework for classification of multi-category hand grasps using EMG signals. Artif. Intell. (2021). https://www.sciencedirect.com/science/article/pii/S0933365720312707
13. Hill-Briggs, F., et al.: Social determinants of health and diabetes: a scientific review. Diabetes (2021). https://www.ncbi.nlm.nih.gov/pmc/articles/PMC7783927/
14. Sun, H., et al.: IDF Diabetes Atlas: global, regional and country-level diabetes prevalence estimates for 2021 and projections for 2045. Diabetes Res. (2022). https://www.sciencedirect.com/science/article/pii/S0168822721004782

15. Ellahham, S.: Artificial intelligence: the future for diabetes care. Am. J. Med. (2020). https://www.sciencedirect.com/science/article/pii/S0002934320303399
16. Cloete, L.: Diabetes mellitus: an overview of the types, symptoms, complications and management. Nurs. Std. (Royal College of Nursing (Great Britain) (2021). https://europepmc.org/article/med/34708622
17. Hasan, M.K., Alam, M.A., Das, D., Hossain, E., Hossain, M.: Diabetes prediction using ensembling of different machine learning classifiers. IEEE Access (2020). https://ieeexplore.ieee.org/abstract/document/9076634/
18. Fagherazzi, G., Ravaud, P.: Digital diabetes: perspectives for diabetes prevention, management and research. Diabetes Metab. (2019). https://www.sciencedirect.com/science/article/pii/S1226236361830171X
19. Romero-Díaz, C., Duarte-Montero, D., Gutiérrez-Romero, S.A., Mendivil, C.O.: Diabetes and bone fragility. Diabetes Therapy (2021). https://doi.org/10.1007/s13300-020-00964-1
20. Saravanan, V., Nivurruti, M., Barde, K., Pillai, A.S., Woungang, I.: Reliable diabetes mellitus forecasting using artificial neural network multilayer perceptron. Artif. Intell. (2022). https://www.sciencedirect.com/science/article/pii/B9780128240540000137
21. Sreedevi, B., Durga Karthik, J., Glory Thephoral, M., Jeya Pandian, G., Revathy, G.: A novel neural network based model for diabetes prediction using multilayer perceptron and Jrip classifier. In: Ranganathan, G., Bestak, Robert, Fernando, Xavier (eds.) Pervasive Computing and Social Networking: Proceedings of ICPCSN 2022, pp. 345–351. Springer, Singapore (2023). https://doi.org/10.1007/978-981-19-2840-6_27
22. Song, H., Lee, S.: Implementation of diabetes incidence prediction using a multilayer perceptron neural network. In: 2021 IEEE International Conference on. ieeexplore.ieee.org (2021). https://ieeexplore.ieee.org/abstract/document/9669583/
23. Polat, S., Parlakpinar, H., Colak, C.: Estimation of the factors associated with diabetes mellitus by multilayer perceptron artificial neural network model. In: Neuroendocrinology. tnedcongress.org (2021). http://www.tnedcongress.org/wp-content/uploads/2020/11/PC-38.pdf
24. Verma, G., Verma, H.: A multilayer perceptron neural network model for predicting diabetes. Int. J. Grid Distrib. (2020). https://www.researchgate.net/profile/Garima-Verma-3/publication/341788367_A_Multilayer_Perceptron_Neural_Network_Model_For_Predicting_Diabetes/links/5ed4ce35299bf1c67d32265d/A-Multilayer-Perceptron-Neural-Network-Model-For-Predicting-Diabetes.pdf
25. Bai, S., Islam, M.R., Power, V., OŚullivan, L.: User-centered development and performance assessment of a modular full-body exoskeleton (AXO-SUIT). Biomimetic Intell. Robot. 2(2), 100032 (2022). https://doi.org/10.1016/j.birob.2021.100032
26. Ferreira, A.C.B.H., et al.: Neural network-based method to stratify people at risk for developing diabetic foot: a support system for health professionals. PLoS ONE 18(7), e0288466 (2023). https://doi.org/10.1371/journal.pone.0288466
27. Islam, M.R.U., Waris, A., Kamavuako, E.N., Bai, S.: A comparative study of motion detection with FMG and sEMG methods for assistive applications. J. Rehabil. Assist. Technol. Eng. 7, 2055668320938588 (2020). https://doi.org/10.1177/2055668320938588
28. Liao, Z., et al.: Human–robot interface based on sEMG envelope signal for the collaborative wearable robot. Biomimetic Intell. Robot., 100079 (2023). https://doi.org/10.1016/j.birob.2022.100079
29. Liu, J., Wang, C., He, B., Li, P., Wu, X.: Metric learning for robust gait phase recognition for a lower limb exoskeleton robot based on sEMG. IEEE Trans. Med. Robot. Bionics 4(2), 472–479 (2022). https://doi.org/10.1109/TMRB.2022.3166543

30. Miften, F.S., Diykh, M., Abdulla, S., Siuly, S., Green, J.H., Deo, R.C.: A new framework for classification of multi-category hand grasps using EMG signals. Artif. Intell. (2021). https://doi.org/10.1016/j.artmed.2020.101884
31. Zhang, J., et al.: Physics-informed deep learning for musculoskeletal modeling: Predicting muscle forces and joint kinematics from surface EMG. Neural Syst. (2022). https://doi.org/10.1109/TNSRE.2022.3226860

Incidence Assessment of Diabetes by Delegation in the United Mexican States Applying the Multilayer Perceptron Neural Network

Hubet Cárdenas-Isla(✉), Rodrigo Leonardo Reyes-Osorio(✉),
Adrián Jacobo-Rojas(✉), Ashlee Robles-Gallegos(✉), and Bogart Yail Márquez(✉)

Instituto Tecnológico de Tijuana Calzada Del Tecnológico S/N, Fraccionamiento Tomas Aquino, 22414 Tijuana, Baja California, México
{m23210003,m23210005,m23210004,ashlee.robles,
bogart}@tectijuana.edu.mx

Abstract. The prevalence and impact of diabetes in Mexico are thoroughly examined in this study. Over the past four decades, diabetes has emerged as the predominant health concern in the country, ranking as the leading cause of death in women and the second in men since 2000. It has also been identified as the primary culprit behind premature retirement, blindness, and kidney failure. Projections indicate that by 2025, nearly 11.7 million Mexicans could be diagnosed with diabetes, underscoring the urgency of understanding and addressing this escalating health crisis [1]. Previous research on diabetes characteristics and consequences among individuals aged 20 to 40 has primarily relied on hospital-based samples, potentially skewing results toward severe cases or specific ethnic groups. A critical gap exists in nationwide, population-based studies that can provide a more comprehensive understanding of the prevalence and characteristics of early-onset type 2 diabetes.

Given that 79% of Mexico's population is under 40 years old [2], there is an imperative need for such studies to inform targeted preventive measures. This study aims to fill this gap by predicting the risk index for the general population based on diabetes incidence data collected by a delegation in Mexico through public health institutions. The prediction will leverage a multilayer perceptron neural network to enhance the accuracy and applicability of the findings.

Keywords: Diabetes · Obesity · Data mining · Neural network · Epidemiology · Ethnic group · Deep Learning

1 Introduction

Health conditions such as diabetes are part of the Mexican population, for which the Mexican health authorities offer a diagnostic program utilizing public institutions to obtain and compile data. Over two decades, Mexico has undergone a remarkable shift in its disease profile, transitioning from a landscape dominated by malnutrition and communicable infectious and parasitic diseases to a nation grappling with the pervasive challenges of obesity [3]. This transformation has significant implications for public health,

Á. Rocha et al. (Eds.): WorldCIST 2024, LNNS 985, pp. 51–59, 2024.
https://doi.org/10.1007/978-3-031-60215-3_6

necessitating a comprehensive exploration of the factors contributing to this shift, focusing on the escalating prevalence of conditions like diabetes. To accurately interpret the data presented in this context, it is imperative to establish a foundational understanding of diabetes. Diabetes is a complex syndrome characterized by liver steatosis, lobular hepatitis, and the persistent elevation of serum alanine aminotransferase (ALT), or non-alcoholic steatohepatitis (NASH) [4]. This multifaceted condition demands a nuanced examination to unravel its intricate manifestations and impacts on public health.

As Mexico grapples with the consequences of this evolving health landscape, it becomes increasingly crucial to delve into the intricate interplay of factors contributing to the surge in obesity-related diseases, with diabetes at the forefront. This study seeks to provide insights into the prevalence and characteristics of diabetes in the country, recognizing the need for a comprehensive understanding of its manifestations and the associated risk factors. By shedding light on the epidemiological landscape of diabetes, we aim to contribute valuable knowledge that can inform targeted interventions and preventive measures, addressing the complex health challenges Mexico faces in the 21st century.

There is a unanimous acknowledgment that type 2 diabetes (T2D) stands as one of Mexico's foremost public health challenges, marked by its soaring prevalence, mortality rates, and morbidity. These stark realities translate into significant human and economic burdens, highlighting the urgent need for comprehensive strategies to address this pervasive issue. Furthermore, a foreboding future looms as the proportion of the elderly population in Mexico undergoes a substantial increase, with T2D manifesting in 25–30% of cases within this age group [5]. Mexico's dietary and physical activity landscape has undergone rapid and profound shifts despite ongoing initiatives promoting exercise habits and restrictions on junk food. From 1998 to 2006, the annual prevalence rate of obesity among Mexican adults surged by approximately 2% per year, ranking among the highest documented globally [3]. National health surveys over the past three decades reveal a consistent uptick in the prevalence of T2D. If these surveys accurately depict the scenario, one plausible explanation is that the rise in prevalence stems from an increase in the incidence of T2D. Consequently, to comprehensively characterize the epidemic's nature, estimating the incidence of this pathological condition becomes paramount.

Drawing parallels with the United States, where the estimated costs attributable to T2D reached 245 billion in 2012, marking a 41% increase from previous calculations, Mexico is grappling with alarmingly high and escalating costs associated with T2D care [5]. This underscores the imperative for a multi-faceted approach that not only addresses the immediate challenges posed by T2D but also considers the long-term implications, emphasizing the urgency of concerted efforts in public health planning and intervention strategies to curb the trajectory of this burgeoning health crisis.

Now, we will focus on the process we use to process the data: Perceptron neural networks (MLP). Multilayer perceptron (MLP) neural networks, characterized by layered structures, play a pivotal role in data processing. These networks consist of units organized into layers, each comprising nodes forming fully connected networks. A standard MLP configuration involves at least three layers: an input layer, one or more hidden layers, and an output layer. It's worth noting that the linear multilayer perceptron, a simplified version with only input and output layers, exists but is typically disregarded.

The input layer distributes inputs to subsequent layers, with input nodes featuring linear activation functions and no thresholds.

Hidden unit nodes and output nodes are equipped with thresholds and weights, with hidden unit nodes employing nonlinear activation functions and outputs utilizing linear activation functions. This intricate process involves multiplying the original input by weight, adding a threshold, and passage through an activation function—linear or nonlinear—when feeding a signal into a node in a subsequent layer [6]. In addition to standard MLP structures, feedback networks can bidirectionally transmit impulses due to reaction connections within the network. These networks, dynamic and potent, continuously evolve until reaching a state of equilibrium. Any change in input triggers a search for a new balance. The introduction of multiple layers was driven by the necessity to enhance the complexity of decision regions [7]. Transitioning to the application of data mining techniques for processing collected data, we turn to the utilization of a multilayer perceptron neural network. This approach facilitates the prediction of the risk index, categorizing it as high, intermediate, or low within the general context of the research.

2 Materials and Methods

The pervasive implementation of artificial intelligence (AI) in the healthcare domain has significantly contributed to augmenting the capabilities of healthcare professionals. This transformative technology extends its utility beyond conventional practices, providing invaluable assistance in decision-making processes. With a particular emphasis on screening classifications and diagnostics, AI systems play a crucial role in enhancing the accuracy and efficiency of medical assessments [8]. Integrating AI-powered tools into healthcare workflows expedites the analysis of vast datasets and introduces a level of precision that is often challenging to achieve through traditional methods. By leveraging advanced algorithms and machine learning models, AI has emerged as a powerful ally, empowering healthcare professionals to make informed decisions, ultimately leading to improved patient outcomes. The continuous evolution and refinement of AI applications in the health sector underscore its potential to revolutionize medical practices and contribute to the ongoing advancements in patient care and diagnosis.

2.1 Healthcare System

In 2001, the Mexican federal government established Seguro Popular (People's Insurance) to safeguard the uninsured from burdensome healthcare costs. Participation in Seguro Popular is voluntary and not contingent on health status or pre-existing conditions. The system operates without co-payments, and contributions are determined solely based on the individual's financial capacity [3]. Despite its acknowledged imperfections, the system's efficacy becomes evident through the data it systematically gathers, employing a predefined fee structure correlated with income deciles. The poorest 4 deciles receive coverage without any payments, while the 5th to the top income deciles contribute amounts ranging from USD 152.00 to USD 834 per family year [9].

Simultaneously, those with economic means opt for medical services from the private sector. Although diabetes mellitus is encompassed within the diseases covered by Seguro Popular, the coverage predominantly extends to ambulatory primary care and urgent medical attention. However, as of 2010, the insurance did not encompass the substantial costs associated with chronic complications like renal failure substitution and acute coronary syndrome. Notably, in 2011, myocardial infarction was incorporated into the catalog of major diseases covered, albeit limited to adults below 60 [12].

2.2 Use of Data with Neural Network

Governments worldwide are increasingly embracing and developing new technologies in the contemporary era, particularly focusing on their practical implementation. Utilizing emerging technologies, notably machine learning and artificial intelligence (ML/AI), presents numerous opportunities to enhance various facets of human life effectively and efficiently. Global governments have proactively incorporated AI methods into their public administration systems, particularly focusing on forecasting methods and predictive analytics grounded in AI/ML technologies [13].

Despite the commendable predictive performance achieved by modern neural networks, there has been ongoing criticism regarding their limited interpretability [15]. This pertains to the challenge of understanding the decision-making process of these networks at a level comprehensible to humans. Often labeled as "black boxes," these networks have difficulty reconstructing the rationale behind their predictions. Therefore, a thorough analysis is essential to uncover the influence of behavior dynamics on prediction performance, establishing data specifications that can serve as benchmarks for evaluating neural networks applied in mobility analysis [14].

The culmination of technological advancements in recent years has spurred considerable research on leveraging artificial intelligence, telemedicine, and mobile health to enhance healthcare outcomes [16]. For this study, we accessed free data from health repositories concentrated in each municipality. Our research does not involve human intervention or collecting biological samples, as we utilize existing databases provided by the National Institute of Public Health and other relevant entities. The data spans from 2000 through 2022, and we employed the multilayer perceptron neural network for analysis, acknowledging the widespread use of artificial intelligence in such applications.

3 Results

The multilayer neural network (MLP) emerges as a supervised learning model that demonstrates its effectiveness using various layers of interconnected nodes or neurons [19]. In the Mexican context, emphasis is placed on the fact that in 2004, a substantial impact of non-communicable diseases was recognized, accounting for 75% of overall deaths and 68% of disability-adjusted life years. Notable contributors to mortality included ischemic heart disease, diabetes mellitus, and cerebrovascular diseases. Factors such as high body mass index (BMI), high blood glucose levels, and alcohol consumption emerged as determinants that contribute substantially to the disease burden, outweighing other risks associated with non-communicable diseases. Diabetes mellitus was positioned as a relevant concern, representing 9.7% of total deaths, with higher rates in women (12.1%) than in men (9.7%) and contributing 3.5% of total DALYs [9]. These findings underscore the urgency of effective and targeted strategies to address public health challenges related to non-communicable diseases, especially diabetes mellitus, in Mexico.

The data were extracted directly from the Mexican Social Security Institute (IMSS) [10, 17]. In the analysis process, a train and test approach were implemented, with an average balance of 70.3% for training and 29.7% for testing, resulting in an average prediction accuracy of 65.4% for the network. Multilayer neuronal (MLP). In detail, this performance translated into an average high-risk level of 21.2%, an intermediate risk level of 17.3%, and a significant low-risk level of 61.5%.

Regarding the configuration of the MLP, the process involves determining the number of hidden layers, specifying the number of nodes in each layer, choosing activation functions (commonly sigmoidal), and defining the training parameters. Each connection between nodes is characterized by a synaptic weight (w) that is iteratively adjusted by an optimization algorithm to minimize the error throughout the training set. This meticulous process ensures efficient adaptation of the MLP to the data set, allowing the network to learn and adjust optimally to achieve the best possible accuracy in predicting the associated risk.

During the operation of a multilayer neural network (MLP), input patterns are initially introduced into the input layer, triggering a layer-by-layer forward propagation process. At each node in the network, nonlinear transformations are applied, and this process is repeated until the predicted outputs are generated at the output layer. Supervised training, facilitated by the error backpropagation method, plays an essential role in allowing the network to iteratively learn to map inputs to desired outputs, progressively adjusting synaptic weights. This continuous and refined adaptation is essential to optimize the predictive performance of MLP. The uniqueness of MLPs lies in their highly effective ability to model complex non-linear relationships between inputs and outputs, making them a robust and powerful approach for predictive analytics and classification tasks in various contexts (Fig. 1).

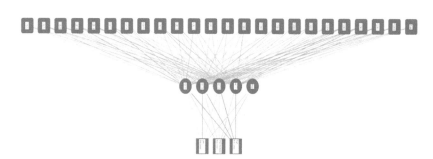

Fig. 1. Multilayer perceptron network diagram with hidden layer hyperbolic tangent activation function [18, 20]

The main mathematical functions involved in the implementation of a Multilayer Perceptron (MLP) neural network i are:

Spread function:

Allows inputs to be propagated from the input layer to the output layer:

$$\text{netj} = \sum \text{iwi} * \text{xi} \tag{1}$$

where xi are the inputs, wi the synaptic weights and netj the net input of node j.

Activation function applies a nonlinear transformation to the net input of each node. For example, the logistics function:

$$\text{yj} = \text{f(netj)} = \frac{1}{\left(1 + e^{(-\text{netj})}\right)} \tag{2}$$

where yj is the output of node j.

Error function:

Calculate the difference between the obtained outputs and the desired outputs to quantify the error made. For example, the mean square error:

$$E = \left(\frac{1}{N}\right) \sum (\text{dj} - \text{yj})2 \tag{3}$$

where N is the number of patterns, dj the desired output and yj the predicted output.

Backpropagation algorithm:

Iteratively adjust synaptic weights to minimize error using gradient descent:

$$\text{wij(new)} = \text{wij(old)} - \eta * \partial E/\partial \text{wij} \tag{4}$$

where η is the learning rate that controls the step of descent.

Weight initialization and optimization functions:

They initialize and update the weights to find the optimal values that minimize the error. SPSS incorporates several algorithms, such as conjugate gradient descent, quasi-Newton, and Levenberg-Marquardt, among others.

In summary, these are the main mathematical functions that model the behavior and learning of MLPs. The software takes care of the computational details, facilitating the use of these powerful neural networks.

In the intricate architecture of this system, the expression of the representation involves W as the parameter or weights within the layer, x as the input vector, which may function as the output of the preceding layer, and b as the bias vector [18, 19] serves as a guiding framework, delineating the fundamental components. It clarifies that W signifies the parameter or weights embedded within the layer, x signifies the input vector, with the added flexibility of serving as the output from the preceding layer, and b symbolizes the bias vector. In this specific scenario, the system is meticulously configured with 23 inputs, spanning the temporal spectrum from 2000 to 2022. The resultant output of the system manifests in three discernible levels: high-risk, intermediate-risk, and low-risk, thereby facilitating a nuanced and all-encompassing risk assessment throughout the stipulated timeframe. This intricate configuration allows meticulous analysis and prediction, capitalizing on the intricate interplay between inputs, weights, and biases intrinsic to the neural network's functionality.

4   Conclusion

A conspicuous trend emerges after meticulously examining the obtained results, revealing that the generated predictions demonstrate a significantly higher average for low-risk scenarios. This intriguing observation instigates a deliberate shift in focus toward both high and low-risk averages, necessitating a nuanced population segmentation strategy tailored to address the complexities of diabetes management in this context. The strategic utilization of the model-derived results empowers us to identify specific demographics precisely, unravel the sources of the disease, and subsequently report these sources to foster a sense of accountability within the healthcare landscape.

Despite ongoing governmental initiatives directed at mitigating the expansion of diabetes-related issues, our data portends a looming escalation in the prevalence and mortality of diabetes in the coming years. This foresight compels a reevaluation of current approaches and underscores the urgency for targeted interventions that can effectively counteract the impending challenges associated with diabetes. The data-driven insights not only serve as a barometer of the effectiveness of existing initiatives but also provide a foundation for recalibrating strategies to address the evolving landscape of diabetes management proactively.

Leveraging the results provided by the model offers a strategic avenue to pinpoint specific demographics, identify the sources of the disease, and subsequently report these findings, thereby fostering a sense of accountability within the healthcare ecosystem. This targeted approach enables a more precise allocation of resources and interventions, tailoring strategies to the unique characteristics of different demographic groups. Despite the government's pre-existing initiatives aimed at combatting the expanding challenge posed by diabetes, our data paints a sobering picture, revealing that the prevalence and mortality of diabetes are projected to persistently escalate in the years to come.

Armed with these results, a targeted approach can be formulated, directing specific programs toward each identified demographic group to prevent the onset of the disease. This entails developing tailored interventions that not only address the unique risk factors associated with each group but also incorporate educational initiatives. By concentrating on education, we can empower and inform populations already affected by the disease, fostering a deeper understanding of its management, and encouraging proactive health measures.

Moreover, ensuring access to essential technologies for both diagnosis and management becomes paramount. This encompasses implementing technological solutions that facilitate early diagnosis, personalized treatment plans, and ongoing monitoring. Equally crucial is the commitment to ensuring the availability of insulin for all individuals who require it, removing barriers to accessibility and prioritizing the well-being of those dependent on this life-saving medication. The comprehensive strategy encompasses a holistic approach that combines prevention, education, and technological advancements to address the multifaceted challenges posed by diabetes, contributing to a more resilient and sustainable healthcare landscape.

Furthermore, recognizing the diverse needs of the population already grappling with the condition, there is an imperative to institute tailored assistance programs. These programs should extend beyond mere medical interventions, encompassing elements that enhance the overall quality of life and promote inclusivity. Such initiatives could include psychological support, community engagement, and accessibility enhancements, all aimed at fostering a more enjoyable and fulfilling life for individuals managing diabetes.

The analysis and systematic compilation of data on the risk levels of the disease assumes a pivotal role in monitoring the impact of the government's ongoing programs designed to combat diabetes. While acknowledging the inherent complexity of such an endeavor, the coordinated and judicious use of this data emerges as a powerful tool in the fight against diabetes. It provides a quantitative measure of program effectiveness and enables iterative improvements and adaptations, ensuring that interventions remain responsive to the evolving landscape of diabetes management. This data-driven approach, when coupled with a commitment to continuous improvement, strengthens the collective resolve to effectively address the challenges posed by diabetes at both individual and societal levels.

Acknowledgment. We sincerely thank the Instituto Tecnológico de Tijuana for their invaluable support and essential contribution to our research as students of the Master's program in Information Technologies. The institution has provided a conducive environment for developing our projects, offering resources and guidance crucial for our academic and professional growth. Additionally, we extend our recognition and thanks to the Instituto Mexicano del Seguro Social (IMSS) for their generous collaboration in providing the necessary information for our research. The availability of accurate and relevant data from IMSS has been instrumental in the success of our work, allowing us to address the challenges posed by our research comprehensively.

We are deeply grateful for the support from both institutions, whose contributions have been pivotal in our academic journey and in achieving our research objectives. This invaluable support has strengthened our foundation as professionals in information technologies, and we are committed to using the knowledge gained to contribute to advancing science and technology. Our gratitude also extends to all those who have participated in this process and made this enriching experience possible.

References

1. Aguilar-Salinas, C.A., et al.: Prevalence and characteristics of early-onset type 2 diabetes in Mexico. Am. J. Med. **113**(7), 569–574 (2002). https://doi.org/10.1016/S0002-9343(02)013 14-1
2. Rull, J.A., Aguilar-Salinas, C.A., Rojas, R., Rios-Torres, J.M., Gómez-Pérez, F.J., Olaiz, G.: Epidemiology of type 2 diabetes in Mexico. Arch. Med. Res. **36**(3), 188–196 (2005). https://doi.org/10.1016/j.arcmed.2005.01.006
3. Barquera, S., Campos-Nonato, I., Aguilar-Salinas, C., Lopez-Ridaura, R., Arredondo, A., Rivera-Dommarco, J.: Diabetes in Mexico: cost and management of diabetes and its complications and challenges for health policy. Glob. Health **9**(1), 3 (2013). https://doi.org/10.1186/1744-8603-9-3
4. Nannipieri, M., et al.: Liver enzymes, the metabolic syndrome, and incident diabetes: the Mexico city diabetes study. Diabetes Care **28**(7), 1757–1762 (2005). https://doi.org/10.2337/diacare.28.7.1757
5. González-Villalpando, C., Dávila-Cervantes, C.A., Zamora-Macorra, M., Trejo-Valdivia, B., González-Villalpando, M.E.: Incidence of type 2 diabetes in Mexico: results of the Mexico city diabetes study after 18 years of follow-up. Salud Pública de México **56**(1), 11–17 (2014)

6. Delashmit, W.H., Manry, M.T.: Recent developments in multilayer perceptron neural net-works. In: Proceedings of the Seventh Annual Memphis Area Engineering and Science Conference, MAESC, pp. 1–15 (2005)
7. Popescu, M.-C., Balas, V.E., Perescu-Popescu, L., Mastorakis, N.: Multilayer perceptron and neural networks. WSEAS Trans. Circuits Syst. **8**(7), 579–588 (2009)
8. Ferreira, A.C.B.H., et al.: Neural network-based method to stratify people at risk for devel-oping diabetic foot: a support system for health professionals. PLoS ONE **18**(7), e0288466 (2023)
9. Villalpando, S., et al.: Prevalence and distribution of type 2 diabetes mellitus in Mexican adult population: a probabilistic survey. Salud Publica Mex. **52**, S19–S26 (2010)
10. IMSS: Detección de Diabetes (2023). por deledación. https://datos.gob.mx/busca/dataset/det eccion-de-diabetes-por-delegacion
11. Barquera, S., et al.: Methodology for the analysis of type 2 diabetes, metabolic syndrome and cardiovascular disease risk indicators in the ENSANUT 2006. Salud Publica Mex. **52**(Suppl 1), S4–S10 (2010)
12. Tobias, M.: Subnational burden of disease studies: Mexico leads the way. PLoS Med. **5**(6), e138 (2008)
13. Ivashchenko, T., Ivashchenko, A., Vasylets, N.: The ways of introducing AI/ML-based predic-tion methods for the improvement of the system of government socio-economic administration in Ukraine. BTP **24**(2), 522–532 (2023)
14. Hong, Y., Xin, Y., Dirmeier, S., Perez-Cruz, F., Raubal, M.: Revealing behavioral impact on mobility prediction networks through causal interventions (2023). ArXiv Preprint arXiv: 2311.11749
15. Manibardo, E.L., Lana, I., Ser, J.D.: Deep learning for road traffic forecasting: does it make a difference? IEEE Trans. Intell. Transp. Syst. **23**(7), 6164–6188 (2022)
16. Alotaibi, M., Aljehane, N.: Early prediction of gestational diabetes using machine learning techniques. J. Theor. Appl. Inf. Technol. **101**(21) (2023)
17. IMSS: istabla43_2022 - Detección padecimientos Diabetes por delegación, por año. 2000–2022 (2023). http://datos.imss.gob.mx/dataset/informacion-en-salud/resource/60f 146be-1528-4abe-8ecc-5daf8f8ca05c
18. Costa, L., et al.: Multilayer perceptron. In: Introduction to Computational Intelligence, vol. 105 (2023)
19. Nitin, KK.: Understanding of Multilayer perceptron (MLP), 21 November 2018. https://med ium.com/@AI_with_Kain/understanding-of-multilayer-perceptron-mlp-8f179c4a135f#:~: text=Each%20layer%20is%20represented%20as,b%20is%20the%20bias%20vector
20. Yao, S.-W., Ullah, N., Rehman, H.U., Hashemi, M.S., Mirzazadeh, M., Inc, M.: Dynamics on novel wave structures of non-linear Schrödinger equation via extended hyperbolic function method. Results Phys. **48**, 106448 (2023). https://doi.org/10.1016/j.rinp.2023.106448

Computer Networks, Mobility and Pervasive Systems

Exploiting Physarum-Inspired Vacant Particle Transport Model to Redesign an Enterprise Network

Sami J. Habib$^{(\boxtimes)}$ and Paulvanna N. Marimuthu

Computer Engineering Department, Kuwait University, Safat, P. O. Box 5969,
13060 Kuwait City, Kuwait
sami.habib@ku.edu.kw

Abstract. Physarum can form a robust network while foraging, which is viewed as a result of microscopic response diffusion followed by a macroscopic spontaneous boundary modification. The vacant-particle transport model apprehends the microscopic-macroscopic responding skills of Physarum to develop a transport mechanism with balanced exploration and exploitation. An existing enterprise network (EN) can be viewed as a set of clusters linking various users, which is analogous to Physarum's nutrition-network that connects a variety of food sources. We propose a redesign process, which explores EN clusters and decides how to exploit them to manage dynamic workloads, as Physarum reforms its network through explore-exploit skills to manage energy for its survival. In this paper, we formulate the redesign process as a two-step optimization problem with an objective function to maximize the managing of workloads at the backbone; here, the 'move user' operation derived from the basic vacant particle model forms the microscopic optimization and the restructuring of EN clusters through evolution forms the macroscopic optimization. A mathematical model with added constraints is developed to expedite the optimization process. The experimental findings demonstrated that the proposed approach reduces the backbone workloads by 39% compared to the initial.

Keywords: enterprise network · microscopic-macroscopic diffusion · redesign · vacant particle transport · optimization · Physarum locomotion

1 Introduction

An enterprise network (EN) can be viewed as a layered network, where the users and the backbone network are distributed at different layers. Enterprise networks are often experienced with a wide variety of workloads, leading to volatile network quality. Sustaining the quality of network may require redesigning existing network infrastructure, which is a challenging task because it involves optimization at multiple levels, such as cost, service, and so on. Thus, an incremental redesign with an objective function to reduce the volume of workloads at the backbone is considered to be a feasible solution to manage the quality of service.

© The Author(s), under exclusive license to Springer Nature Switzerland AG 2024
Á. Rocha et al. (Eds.): WorldCIST 2024, LNNS 985, pp. 63–72, 2024.
https://doi.org/10.1007/978-3-031-60215-3_7

Our prior works on network redesign were based on a number of custom-made tools, where the redesign schemes generated alternate topologies by relocating the users randomly within EN clusters. The bio-inspired algorithms, such as Genetic Algorithm [1, 2], Ant-Colony Optimization [3], Molecular Assembly [4, 5], Tabu Search [6] and Simulated Annealing [7] were employed to explore the search space comprising of alternate topologies. In a recent work on redesign [8], Physarum's classical shortest-path algorithm is modified to group a pair of users, where one of them has appeared more frequently in the set of shortest paths that have been generated from the other to all remaining users considered as destinations.

Physarum, an acellular, multinucleate amoeboid, exhibits a balanced exploratory and exploratory phases. It expands its tubular vein structure in random directions to scavenge for food sources in its exploratory phase; the tubes consist of an Actin-Myosin fibril wall (cytoskeleton 'gel') surrounding a cytoplasmic 'sol' (liquid) [10]. When a quality food source is sensed by the receptor molecules on the tube's boundary surface, the protoplasmic streaming frequency increases in the vicinity of the food source, changing the internal hydrostatic pressure, and causing variations in cytoplasmic flow. Locally transported particles trigger a sol-gel transformation, and the boundary conditions are constantly changing in proportion to the volume of particles transported [11]. Physarum extends its search without disconnecting the explored food sources with sufficient nutrients while conserving the total protoplasmic mass.

Gunji et al. [12] developed a vacant particle transportation model (VPTM) assimilating Physarum's spontaneous chain reaction of protoplasmic flow, where he considered four states to simulate the model: vacant particle, vacant state, sol, and gel. The model simulated aggregated biomass as aggregated sol-state cells surrounded by cells in vacant-state. The transportation of a vacant particle from the set of vacant-sate cells is implemented through a series of actions: random selection of a vacant particle, movement of the vacant particle to its sol-state cells neighborhood at random, followed by changing its state to gel-state. Once it reached the gel-state, it cannot be transported further. This procedure continues until all of vacant-state cells have transformed to gel-state. Then, all the gel-state cells are transformed into sol states, and the entire process is repeated until the vacant cells move out of their neighborhood or the required number of iterations is exceeded. The number of cells composing the aggregated mass is conserved during the transportation.

In VPTM, a system is viewed as a collection of particles, which are capable of moving and transforming themselves as well as the system's shape. Inspired by the VPTM [12], we have developed a redesign process with an objective function to manage the workloads at the backbone. Our contribution is threefold: i) modification of existing vacant particle transportation model to suit to EN redesign problem, ii) formulation of association matrix and additional constraints to speed up the optimization process, and iii) derivation of mathematical model to simulate the redesign process.

In this paper, we have formulated the redesign process as a two-step optimization process: i) relocation (microscopic), and ii) adaptation (macroscopic); transporting a user from an individual cluster and observing the change in workload is viewed as a microscopic reaction, and subsequent updation of workloads at the backbone is viewed as a macroscopic (global) transformation. In similarity to VPTM, the enterprise network is perceived as a group of users that are capable of moving among the clusters, thus transforming EN structure. We have assigned four states, such as {*receptive*(r), *dynamic*(d), *static*(s), *transitive*(t)} to EN users in analogy with sol, vacant, gel and vacant particle respectively in paper [11]. A set of randomly generated clusters possessing r-state users form the aggregated cells and the vacant particle movement is implemented through random user selection followed by relocation. A cluster producing a large volume of outgoing workloads is selected as the source cluster (analogous to vacant–state cells) and its *r-state* users are changed to *d-state*. A user selected randomly from the group of *d-state* users is designated as *t*- and is capable of moving to one of its associated clusters, selected as the destination cluster (analogous to neighborhood sol-state cells); the user-cluster associations are derived using an association factor, which represents the total weighted association between the selected user and the cluster. The movement causes a state change from *t*- to *s*-. This set of procedures constitutes the microscopic process, and it is repeated until the source cluster is left with the defined minimum number of users. Constraints are added for the selection of destination clusters, as well as user relocation to suit to our problem domain. In the adaptation step, the *s-state* cells are reset to the *r-state* and the present enterprise network clusters with modified user distribution form the network to redesign. The entire process is repeated until there is no significant change in the backbone workloads or the exceeding of the specified number of iteration. The experimental results demonstrated that the proposed redesign process reduced the backbone workload by 39% than the initial.

2 Background

Our extensive literature survey has revealed that the research on Physarum either focused on biological aspects or on bio-inspired applications. Many biological experiments concentrated on studying the exploratory and exploitation behaviors under varying environmental conditions, the ability to connect food sources, the embedded intelligence to survive and learn, and the evolutionary behavior to form an optimum global network structure [13–16]. The bio-inspired applications focused on simulating Physarum's intelligence behavior and in solving complex real-world problems. The bio-inspired applications have become popular since the experiments carried out by Nakagaki et al. [17] to demonstrate the shortest path through a maze. Subsequent improvements have confirmed the capability of Physarum to solve a wide range of problems, such as decision-making [18, 19], shortest path finding [20], spatial optimization [21, 22] and so on.

In this work, our focus is on biological experimental outcomes related to Physarum's locomotion and the impact of food quality on its exploration [23, 24]. According to the biological studies by Gunji et al. [9], an adaptive and robust network creation was made by microscopic response diffusion and macroscopic spontaneous boundary modification. A subsequent study by Latty and Beekman [25] on how food quality impacts Physarum's behavior revealed that Physarum plays an important role in integrating and processing sensory and nutritional information in order to select a food source to exploit, the time to spend on exploiting the discovered food sources, and the time and place to resume the exploration. Further, the authors also investigated the influence of nutrient quality on altering the vein structure and morphology—shorter and thicker or longer and thinner veins in response to food quality [26]. Gunji et al. [12] developed a vacant particle transportation model to simulate Physarum's locomotion and compared it with the real-time behavior patterns obtained from laboratory tests under controlled environments. The research comprised three model developments: the transportation model, the shrinkage model, and the expansion model. Later on, Liu et al. [27] designed a transport network in Sichuan province of China based on a vacant particle model with shrinkage and compared the performance with a real-world transport network. A recent research by Liu et al. [28] improved the work in paper [27] by proposing an agent-based evolution model to simulate Physarum's foraging model, where the authors claimed that the developed transport network model showed better performance, such as a shorter average length and higher efficiency than a real-world network.

In this paper, we followed the basic vacant particle transportation model proposed by Gunji et al. [12] to develop a redesign process with added constraints and modifications to suit our problem domain.

3 Modeling Enterprise Network Redesign Through Vacant Particle Transportation Model

In Physarum, the spontaneous transformation of gel and sol change the thickness of the tubular structure, facilitating the ability to alter the existing mesh network structure. Sol is a protoplasmic fluid, whereas gel is a sponge-like structure supporting the membrane of the tube. If the membrane collapses due to a local environmental changes, surrounding sol flows to the site and it is then transformed into gel to support the membrane. This chain reaction causes a spontaneous protoplasmic flow, which is utilized by Gunji et al. [12] to develop the VPTM.

We have formulated the enterprise redesign problem as an optimization problem to maximize volume of workloads locally, as demonstrated in Eq. (1), where the term $\delta_{i,j}$ is a binary variable and it takes 1 if there is an edge between users i and j within the selected cluster; the volume of workload through each edge is represented by the term $tr_{i,j}$. We have added constraints (2) to (8) to facilitate controlled cluster-formation, where, the constraints for user distribution within an individual cluster at any instance, the source and destination cluster selection and the law of conservation are presented respectively. The user movement is restricted if the relocation results in violation of threshold limits of cluster as in constraint (3) and (4); moreover, the cluster with full of s-state users is eliminated in the source clusters' selection list. We have defined a matrix called association matrix (A), such that its element $a_{ij} = w$, if $n_i \to n_j$; Otherwise zero. Here, the term w is the weight factor, which represents the volume of workload. The diagonal elements are made zero to avoid self-loops. The user transportation from a source to a destination cluster is facilitated if the destination possess any of the associated users of the user to be transported as explained in constraint (6). The total number of users within the clusters are conserved as in Constraint (5).

$$z = \max \sum_{j=1}^{N-1} tr_{n_i, n_j} * \delta_{i,j} \quad n_i \in c_i \& n_j \in c_i; i \neq j \tag{1}$$

Subject to:

$$C_L \leq \sum |c_i| \leq C_U \tag{2}$$

$$\forall c_i \in C, \exists c_i | c_i \neg \in c^{so}, \text{ if } |c_i| == C_L \tag{3}$$

$$\forall c_i \in C, \exists c_i | c_i \neg \in c^{dn}, \text{ if } |c_i| == C_U \tag{4}$$

$$\sum_{c_i \in C} |c_i| = N \tag{5}$$

$$t \rightarrow c_i | c_i \in C_t^A \tag{6}$$

We made an assumption that a user can take any of the three states based on its current state, as defined as in Eq. (7). A source cluster $c_i^{so} \in C$, is chosen from existing K clusters that producing higher extra-volume of workloads and the state of all *r-state* users in the source cluster are changed to the *d-state*, which is analogous to the sol-state cells surrounded by vacant-state cells in VPTM. Movement and state change are restricted for *s-state* users during the microscopic process.

$$\forall n_i \in N, St(n_i) = \begin{matrix} r & (n_i \in c_i^{dn} \ \& \ St(n_i) \neq s) \\ d & (n_i \in c_i^{so} \ \& \ St(n_i) \neq s) \\ s & (n_i^d \rightarrow n_i^t) \end{matrix} \tag{7}$$

Since our objective function is to maximize intra-cluster workloads in the existing enterprise network (EN), we did not consider distance as a parameter to assign the next location to move to, as in the paper [12]. Instead, we have defined a factor called association factor (*AF*) of a user n_i to a cluster c_i, as shown in Eq. (8), which is used in the selection of a list of clusters as a preferred locations to move; the clusters' that possess one or more direct connected pairings of the selected user n_i are the destination clusters c_i^{dn}. The denominator of Eq. (8) reflects the total degree of the user n_i, whereas the numerator term represents the number of edges of n_i in cluster c_i, and the term e_{n_i,n_j} takes the values [0, 1] to indicate the presence of an associated user in the chosen cluster.

$$AF(n_i, c_i) = \frac{\sum_{\forall n_j \in c_i} e_{n_i,n_j}}{\sum_{k=1}^{K} \sum_{\forall n_j \in c_k} e_{n_i,n_j}} \tag{8}$$

4 Computational Flow of the Proposed Redesign Process

We have generated a set of clusters with randomly distributed users as an analogy to the aggregated cells in VPTM, where the users are set to r-state (sol state). The flowchart of the redesign process is demonstrated as in Fig. 1. Each user in the given EN is represented by 5-tuples, as n = <c, id, LN, St, AN>, where c is the associated cluster index of the user, id is the user index, LN is the list of workloads from all users, St is the state the user is currently in and AN is the associated users list. Here, the user is in any of the four states as represented by St = {r, d, t, s}. We have used lower and upper thresholds for user distribution to avoid overcrowding of users in a single cluster. The cluster producing higher volume of outgoing workload is selected as the source cluster, where the edges of users' contributing low to the intra-cluster transactions are to be moved, as analogous to moving the vacant particle out of the aggregated cells [12, 27].

Figure 1 explains the overview of the redesign process, where the main flow path explains the cluster selection and transportation of user from the cluster to an associated cluster using the exploited VPTM (microscopic). Then, it explains the adaptation process as analogous to gel to sol breakout in VPTM, where all users states are changed to r-state. Subsequently, the redesigned clusters form the basic network, and the entire process of

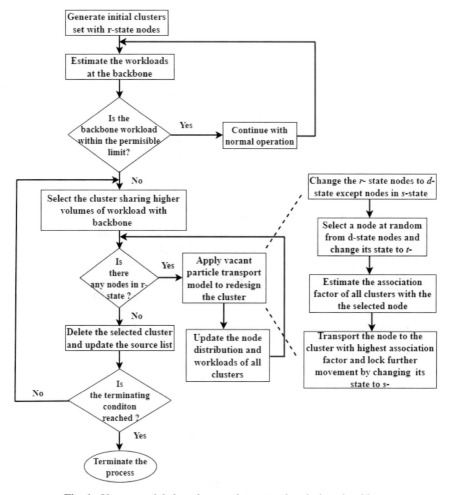

Fig. 1. Vacant particle based enterprise network redesign algorithm.

selection and relocation are repeated till there is no change in the backbone workloads or the specified number of iteration is exceeded (macroscopic).

The user transportation process within EN is explained in the subsection, where the selected cluster users are transformed to d-state (vacant state), and a randomly selected user from the group of d-state users is changed to t-state (vacant particle). The transport of the selected t-state user is implemented by moving the user from the selected cluster to a cluster with more association factor values and change to s-state to avoid repeated selection of the same user. The local transactions of each cluster are updated after every user movement, and then, d-state users are changed to r-state and process is repeated until the source cluster size reaches the lower limit.

5   Results and Discussion

The redesign process based on VPTM is coded in Python platform. We evaluated the performance of the evolved EN by quantifying the volume of workloads at the backbone. In order to get comprehensive experimental results, the developed algorithm is run multiple times on a medium-sized EN, where EN users are distributed in m clusters as specified by the user, and the initial EN is considered as a single cluster. The number of clusters and the total number of users distributed remained constant throughout the experiment, supporting the conservation of mass of protoplasm during the exploration and exploitation phases in Physarum.

The user movement from one cluster to another is accomplished through a chain reaction, where the users undergo state changes from $r \rightarrow d \rightarrow t \rightarrow s$. The user movement decreases the number of users in the selected source cluster while increasing the number of users in the destination cluster, simulating a conserved protoplasmic mass flow. In VPTM, a balance between exploitation and exploration is maintained by using path-source economy (exploration) and minimal path length to improve transport efficiency (exploitation). In our algorithm, the tradeoff between exploration and exploitation is maintained by introducing two parameters: the inter-cluster workload volume during exploration and the association factor when transporting the user. The cluster patterns that get generated during the optimization depend on the previous user distribution and cluster size threshold parameters, limiting user movement across the clusters. The experiment is repeated a maximum of $n - l$ times to transform all *d-state* users to *s-state,* where l is the minimum number of users that should be maintained by any cluster.

5.1   Behavior of the Proposed Redesign Algorithm

We have considered an EN with 20 users generating an initial traffic of 133 MB. The number of clusters is limited to four, and the lower and upper threshold of users distributed within each cluster are two and seven respectively. A set of four clusters with randomly distributed users in *r-state* formed the initial solution; the extra-traffic from each cluster is estimated and the cluster showing a maximum extra-traffic is selected as the source cluster. A state change from *r-* to *d-* is followed by a random selection of transporting user t among *d-state* users in the source cluster; constraints are added to restrict the movement from the source cluster if the movement violates the cluster size thresholds or destination clusters are unavailable, where only state change is allowed. A factor called association factor as defined in Sect. 3 is utilized to select the destination clusters.

The backbone workload, which is the sum of the weighted transactions out of each cluster to the remaining clusters, is calculated for each move, and a snapshot of backbone

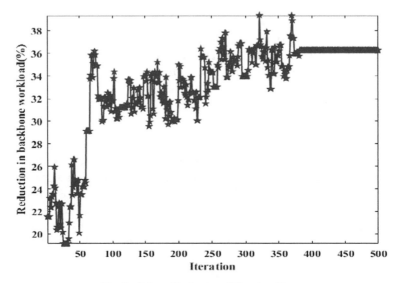

Fig. 2. Internal behavior of the algorithm.

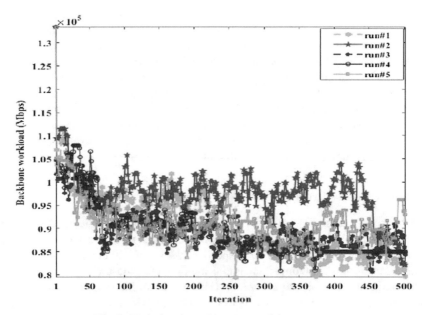

Fig. 3. Redesign through vacant particle transport.

workload variations is demonstrated as in Fig. 2. Here, we have observed an average of 32.27% reduction in backbone workload from the single clustered EN and an average of 15.6% reduction against the backbone workload than the initial random clustered EN. The experiment is repeated multiple times, and a sample of five iterations is presented in Fig. 3. It is observed that the backbone workload reduction slightly varies according to the initial user distribution, and the reduction varies as 40.5%, 38.3%, 39.5%, 39.38%, and 38.1%, respectively, for runs #1 to #5. On an average, the backbone workload is reduced by 39.15% than the existing EN.

6 Conclusion

We have formulated the enterprise network redesign problem as a microscopic-macroscopic optimization problem with an objective function to maximize the intra-cluster workloads, where the redesign process evolved through a modified vacant-particle transportation model. EN is viewed as a collection of users that are capable of moving and we have assigned four states to the users as in VPTM, such as r-,d-,t-, and s-, to relocate the user. The redesign process is viewed as exploration-exploitation process, where transporting a user across the clusters grow the cluster producing lower volume of workloads through the backbone. We have derived an association factor to find best destination cluster so as to expedite optimization. The experimental results on a medium-sized EN demonstrated that the redesign process reduced the backbone workload volumes by 39% than the initial.

Acknowledgement. This work is supported by the Research Sector, Kuwait University, under a research grant no. EO08/19.

References

1. Habib, S.J., Marimuthu, P.N., Taha, M.: Network consolidation through soft computing. In: The Proceedings of International Symposium on Methodologies for Intelligent Systems, Prague, Czech, 14–17 September 2009
2. Habib, S.J., Marimuthu, P.N., Zaeri, N.: Carbon-aware enterprise network through redesign. Comput. J. **58**(2), 234–245 (2015)
3. Habib, S.J., Marimuthu, P.N.: A bio-inspired tool for managing resilience in enterprise networks with embedded intelligent formulation. Expert. Syst. **35**(1), 1–14 (2018)
4. Habib, S.J., Marimuthu, P.N.: Self-organization in ambient networks through molecular assembly. J. Ambient. Intell. Humaniz. Comput. **2**, 165–173 (2011)
5. Habib, S.J., Marimuthu, P.N., Hussain, T.H.: Enterprise network sustainability through bio-inspired scheme. In: The Proceedings of International Conference on Bio-Inspired Computing - Theories and Applications, 16–19 October, Wuhan, China (2014)
6. Hussain, T.H., Marimuthu, P.N., Habib, S.J.: Exploration of storage architectures for enterprise network. Comput. J. **61**(2), 233–247 (2018)
7. Habib, S.J., Marimuthu, P.N.: Optimization of network parameters through redesign operations within simulated annealing. Kuwait J. Sci. Eng. **38**(1(B)), 167–190 (2011)

8. Habib, S.J., Marimuthu, P.N.: Physarum inspired enterprise network redesign. In: The Proceedings of World Conference on Information Systems and Technologies, 4–6 April, Pisa, Italy (2023). https://doi.org/10.1007/978-3-031-45642-8_1
9. Gunji, Y.-P., Shirakawa, T., Niizato, T., Haruna, T.: Minimal model of a cell connecting amoebic motion and adaptive transport networks. J. Theor. Biol. **253**(4), 659–667 (2008)
10. Pollard, T.D., Borisy, G.G.: Cellular motility driven by assembly and disassembly of actin filaments. Cell **112**(4), 453–465 (2003)
11. Reid, C.R., Latty, T., Dussutour, A., Beekman, M.: Slime mold uses an externalized spatial "memory" to navigate in complex environments. Proc. Nat. Acad. Sci. **109**(43), 17490–17494 (2012)
12. Gunji, Y.-P., Shirakawa, T., Niizato, T., Yamachiyo, M., Tani, I.: An adaptive and robust biological network based on the vacant-particle transportation model. J. Theor. Biol. **272**(1), 187–200 (2011)
13. Alim, K., Andrew, N., Pringle, A., Brenner, M.P.: Mechanism of signal propagation in Physarum Polycephalum. Proc. Nat. Acad. Sci. USA **114**(20), 5136–5141 (2017)
14. Christina, O., Nakagaki, T., Dobereiner, H.-G.: Slime mold on the rise: the physics of Physarum Polycephalum. J. Phys. D Appl. Phys. **53**(31), 310201–310210 (2020)
15. Werner, L.C.: Disruptive material intelligence of Physarum: liquid architecture of a biological geometry computer. In: Adamatzky, A. (ed.) Slime Mould in Art and Architecture, pp. 227–247. River Publishers, Gistrup (2019)
16. Reid, C.-R., Beekman, M.: Solving the Towers of Hanoi - how an amoeboid organism efficiently constructs transport networks. J. Exp. Biol. **216**(9), 1546–1551 (2013)
17. Nakagaki, T., Yamada, H., Tóth, Á.: Maze-solving by an amoeboid organism. Nature **407**(5964), 439–442 (2000)
18. Iwayama, K., Zhu, L., Hirata, Y., Aono, M., Hara, M., Aihara, K.: Decision-making ability of Physarum Polycephalum enhanced by its co-ordinated spatio-temporal oscillatory dynamics. Bioinspir. Biomim. **11**(3), 1–15 (2016)
19. Gao, C., et al.: Does being multi-headed make you better at solving problems? A survey of Physarum-based models and computations. Phys. Life Rev. **29**, 1–26 (2019)
20. Zhang, X., Zhang, Y., Zhang, Z., Mahadevan, S., Adamatzky, A., Deng, Y.: Rapid Physarum algorithm for shortest path problem. Appl. Soft Comput. **23**, 19–26 (2014)
21. Dhawale, D., Kamboj, V.K., Anand, P.: An effective solution to numerical and multi-disciplinary design optimization problems using chaotic slime mold algorithm. Eng. Comput. **38**(4), 2739–2777 (2022)
22. Zhu, L., Aono, M., Kim, S.J., Hara, M.: Amoeba-based computing for traveling salesman problem: long-term correlations between spatially separated individual cells of Physarum Polycephalum. Biosystems **112**(1), 1–10 (2013)
23. Ueda, T., Hirose, T., Kobatake, Y.: Membrane biophysics of chemoreception and taxis in the plasmodium of Physarum Polycephalum. Biophys. Chem. **11**(3–4), 461–473 (1980)
24. Cavender, J.: Myxomycetes: a handbook of slime molds. Bioscience **45**(11), 795–797 (1995)
25. Latty, T., Beekman, M.: Food quality affects search strategy in the Acellular Slime-mould. Physarum Polycephalum Behav. Ecol. **20**(6), 1160–1167 (2009)
26. Latty, T., Beekman, M.: Irrational decision-making in an amoeboid organism: transitivity and context-dependent preferences. Proc. Biol. Sci. **278**(1703), 307–312 (2011)
27. Liu, Y., Gao, C., Liang, M., Tao, L., Zhang, Z.: A Physarum-inspired vacant-particle model with shrinkage for transport network design. In: The Proceedings of International Conference on Swarm Intelligence, 25–28 June, Beijing, China (2015)
28. Liu, Y., Gao, C., Zhang, Z.: Simulating transport networks with a Physarum Foraging model. IEEE Access **7**, 23725–23739 (2019)

Simulation of Autonomous Driving Systems for the City of Macas Using Matlab Driving Scenario Design App

Juan Espinoza Gonzalez and Edgar González(✉)

Universidad Tecnológica Israel, E4-142, Marieta De Veintimilla y Fco. Pizarro, Quito, Ecuador
{e1400799852,eegonzalez}@uisrael.edu.ec

Abstract. This paper refers to the use of the Driving Scenario Designer App tool of Matlab, in a virtual simulation environment of autonomous driving systems applied to the city of Macas, Morona Santiago Province, Ecuador. The objective of this research is to develop scenario simulations to obtain information about the object detection of an automatic driving system for cars at the intersections of Av. 29 de mayo and Gavino Rivadeneira in the city of Macas. The results obtained from the three scenarios proposed, whose arrangement of sensors is based on the Waymo One electric car whose configuration is based on three types of camera, radar and ultrasonic sensors, obtained the best detection results.

Keywords: Matlab · ADAS · DSD · Automated Driving toolbox · Detection · Autos autonomous

1 Introduction

According to [1] in his article: A Credibility Assessment Approach for Scenario-Based Virtual Testing of Self-Mapped Driving Functions, he states that dynamic obstacles are the most significant for accidents causing strong and unfavorable claims for the driver. Furthermore, in [2] in his article "Intelligent vehicle impact characterization and accident notification system", he states that vehicle accidents are mainly caused by driver problems due to not paying attention to the inadequacies of roads and highways.

In reference to [3], in the statistics of traffic accidents, in the province of Morona - Santiago Ecuador 136 traffic accidents were recorded for the year 2021, for the year 2020 121 accidents were recorded, the causes of accidents are mainly due to driver recklessness and pedestrian recklessness, in addition to the percentage of vehicles involved in these accidents are motorcycles with 22% and cars with 21.7%. Therefore, ADAS systems (Advanced Driver Assistance System) is one of the rising trend and presents one of the proposals and challenges of today, since they must take care of safety, for that it is essential that vehicles are electric accompanied by sensors for such special assistance. These automated systems nowadays take advantage of technology such as IOTs and artificial intelligence for their respective communications, in addition to making the best driving decisions [4].

Á. Rocha et al. (Eds.): WorldCIST 2024, LNNS 985, pp. 73–81, 2024.
https://doi.org/10.1007/978-3-031-60215-3_8

This document is organized as follows: Section 2 describes the research methodology. Section 3 briefly describes the development of the research. Section 4 describes the results found in accordance with the research objective. Section 5 shows the discussion of the research results. Section 6 presents the conclusions.

2 Methodology

The research approach is quantitative, since numerical results of the respective simulation will be generated and a descriptive analysis will be made showing the best simulation option. The type of research is non-experimental, transversal because a simulation of the automobiles applied to Macas Morona Santiago will be carried out. Finally, the research technique to be used in this document is Driving Designer App of Matlab [5], where the real scenario of the city of Macas is generated with their respective approaches applying the theory of advanced driver assistance systems ADAS, with the respective appropriate algorithms to generate such scenarios and load the relevant sensors so that they can detect the respective data.

After the creation of the scenario, simulation of each of the agents, the data is exported with the respective detections, and the respective script is made to obtain the detections of the sensors and the number of detections for each object or agent, finally each of the variables mentioned above is shown in graphs for their respective interpretation.

3 Development

For the simulation, three respective scenarios were taken according to the information obtained from traffic accidents and according to the base autonomous vehicles available on the market. The respective scenarios are detailed below [6]:

Scenario 1: The following actors are proposed: an ego-type car, a pedestrian, a cyclist and a trailer. The car follows a trajectory (see Fig. 1) and the other actors are static.

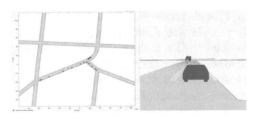

Fig. 1. Scenario 1, for the city of Macas

Scenario 2: The following actors were incorporated: two ego cars, a trailer, a pedestrian and a cyclist. The trailer and the two cars are in motion with their respective trajectory (see Fig. 2) and the cyclist is static.

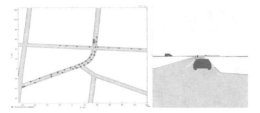

Fig. 2. Scenario 2 for the city of Macas

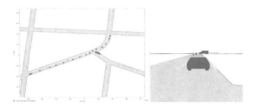

Fig. 3. Scenario 3 for the city of Macas

Scenario 3: The following actors are implemented: an ego car, a trailer, a pedestrian and a cyclist. The ego car with its particular trajectory (see Fig. 3) and the cyclist, pedestrian and trailer in a static way.

The configuration of the respective pathways, actors and sensors are described below. *Roads and Location*: To perform the respective configuration of the roads, within the DSD (Driver Scenario Designer) app of Matlab, the Open StretMap complement was used, where the location of the streets, which are Av. 29 de mayo and Gavino Rivadeneira street, was georeferenced. Below is the OpenStretMap environment (see Fig. 4) with the respective import of the intersection of interest.

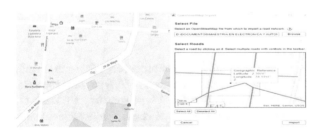

Fig. 4. Location from Open StretMap

Actors and Trajectories: Within the DSD app, the respective configuration of the actors used [7], a pedestrian, a trailer and cars, as appropriate, was performed. In this configuration, the location of each of them with their respective trajectory was made, this according to the information provided by the national traffic agency and their accidents (see Fig. 5).

Fig. 5. Simulation in Unreal - Scenarios

Sensors: For the configuration of the sensors we proceeded to use the auto ego provided by DSD, and the sensor was configured with the respective dimensions according to the car that is on the market. Each scenario is described below.

Sensors for Scenario 1
A WAYMO ONE type of electric car was proposed, with a configuration of three types of sensors, where 4 camera type sensors are located 2 in front and 2 on the sides of the back of the car, in addition to a Radar type sensor in the front, with a range of 75 m, and finally an ultrasonic sensor, with a range of 12 m (see Fig. 6).

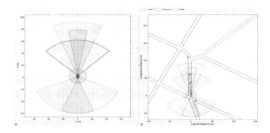

Fig. 6. Location of Sensors - Scenario 1

Sensors for Scenario 2
An electric car type TESLA Model S was proposed, where the configuration of the sensors was taken as a basis, which are composed of 4 types of sensors, these are LiDAR, Camera, Radar and Ultrasonic. Four camera type sensors, two radar type sensors, one ultrasonic type sensor and one LiDAR type sensor were implemented. For both the camera and radar type sensors a range of 25 m was implemented, with a focal length of (320 m, 320 m) for the two side sensors of the car and for the front and rear sensors of (800 m, 800 m). For the Ultrasonic sensor the range is 8 m, and its AZimuth field of view is 70°, for the LiDAR type sensor it is 20 m, and its angle of view is 360° (see Fig. 7).

Sensors for Scenario 3
The Lexus LS 500 electric car was proposed, where the base configuration of the sensors has been taken in the same way, implementing three types of sensors, camera type, radar type and LiDAR type. Four camera type sensors were placed, two lateral and two on the

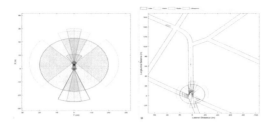

Fig. 7. Location of Sensors - Scenario 2

front and rear of the car, the radar type sensors were placed three, one on the front and two on the side of the car. Their respective position height field of view (see Fig. 8).

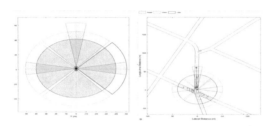

Fig. 8. Location of Sensors - Scenario 3

4   Results

The results corresponding to each of the respective scenario simulations are presented below.

Simulation Scenario 1
For the simulation of scenario 1, the following simulation is carried out with the Driving Scenario Designer (DSD). As can be seen in the DSD simulation, there are three respective views: the first is the egocentric view, in which the simulation is shown in the form of cubes of the respective actors. The bird's eye view is the one that shows the sensors with their respective detection and finally, it shows the simulation of the space where the main road is located. In Fig. 9. The detection of objects is shown.

Simulation Scenario 2
For scenario 2, three simulation views are shown in the same way as for scenario 1. Figure 10 shows the trajectory and sensor detections respectively.

Simulation Scenario 3
Finally for Scenario 3, the following results were obtained in DSD, as in the previous simulations, three views are presented. The following Fig. 11 shows the respective trajectory, movement for the actors and their detection of objects from the respective sensors.

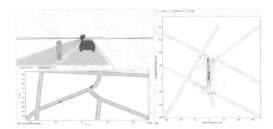

Fig. 9. Simulation in DSD - Scenario 1, city of Macas

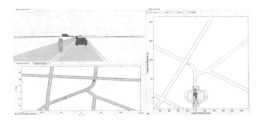

Fig. 10. Simulation in DSD - Scenario 2, city of Macas

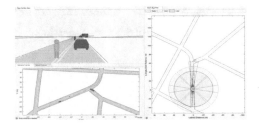

Fig. 11. Simulation in DSD - Scenario 3, city of Macas

Algorithm Generation

Once each of the respective simulations has been generated, with the MATLAB function command, we proceed to analyze the respective data in order to extract the detection of the sensors and their respective positions.

For vehicle position, speed and acceleration, the following functions are shown below

$$Positionx(i) = data(i).ActorPoses(1).Position(1) \qquad (1)$$

$$Positiony(i) = data(i).ActorPoses(1).Position(2) \qquad (2)$$

$$velocityx(i) = data(i).ActorPoses(1).Velocity(1) \qquad (3)$$

$$velocityy(i) = data(i).ActorPoses(1).Velocity(2) \qquad (4)$$

$$accx = \left[diff(velocityx)0 \right]; \; accy = \left[diff(velocityy)0 \right] \qquad (5)$$

And finally for the extraction of the number of detections and detection measurements of the sensors is shown below.

$$Detection(i, k + 1) = data(i).ObjectDetections\{n, 1\}.Measurement(k) \qquad (6)$$

$$NumDetection(i) = length(data(i).ObjectDetections) \qquad (7)$$

5   Discussion

At this point we will show the results of the simulation created in DSD, we worked on three scripts, where it contains code and shows the speed, acceleration, detection measurements and the number of detections. Then the respective results will be shown per scenario.

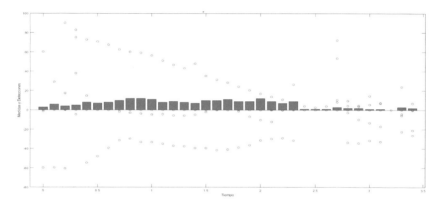

Fig. 13. Measurements and Detections - Scenario 1, for city of Macas

Result Scenario 1
Figure 13 shows the number of detections and the detected measurements, showing the detections in a wide time period from 0 to 2.3 with a frequency of 12 detections of objects in the area. In addition, the range for object detection is 90 m, having low levels of detections and number of detections in the range of 2.4 to 3.5.

Result Scenario 2
Similarly, Fig. 14 shows the detection of objects with the respective measurements, and the interval with the highest proportion of detections is from time 2 to 2.5, with a frequency of 8 detections respectively. Furthermore, it can be differentiated that the time grows with positive trend, the measurements are produced according to the respective number of detections.

Result Scenario 3
Figure 15 shows the sensor measurements and the number of detections, which agree in time between 0–2.3 and 2.5–3.8 respectively. In addition, the maximum frequency of detections is 12, with a consideration and a range of 80 m for a detected object.

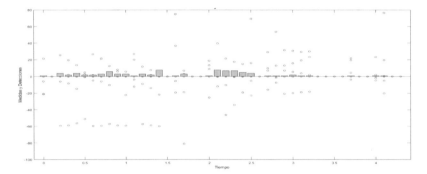

Fig. 14. Measurements and Detections - Scenario 2, for city of Macas

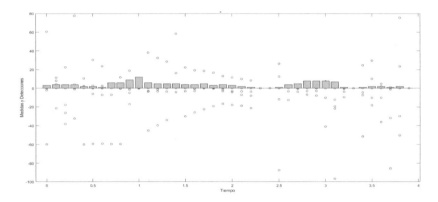

Fig. 15. Measurements and Detections - Scenario 3, for city of Macas

6 Conclusions

Based on the simulations of the three scenarios and according to the graphs presented, the following conclusions can be drawn:

- For the simulation, 3 scenarios were proposed respectively, based on a configuration of base sensors of three manufacturers of autonomous electric cars such as Waymo, Tesla and Lexus, where they were taken as a basis for the configuration of sensors, taking the height, angle of aperture and range of the sensors respectively. Once these scenarios were configured, the Driving Scenario Designer App of MATLAB was applied to observe the behavior of the sensors and their measurements, in addition to the respective speeds and acceleration in the city of Macas, Morona Santiago, Ecuador.
- According to the results found in the simulation of scenario 2, where 4 types of sensors were configured and the agents are in a dynamic situation, it is concluded that the detection of the objects is more complex, giving way to an in-depth analysis of dynamic situations and their behavior for future research.

- The sensor configuration that produces the greatest detection and range according to the time between 0–2.4 detection is for scenario 1, where a base scenario of the WAYMO ONE car is configured, whose sensors are placed at a height of 2 m for the front cameras and 0.80 m for the radar and the ultrasonic sensor. The range given to this configuration is 90 m, 60 m for the cameras, 75 m for the radar and 12 m for the ultrasonic sensor, finally the detection frequency is 12 with the objects proposed in the scenario.

References

1. Stadler, C., Montanari, F., Baron, W., Sippl, C., Djanatliev, A.: A credibility assessment approach for scenario-based virtual testing of automated driving functions. IEEE Open J. Intell. Transp. Syst. **3**, 45–60 (2022). https://doi.org/10.1109/ojits.2022.3140493
2. Navin Kumar, M., Pravin Kumar, S., Premkumar, R., Navaneethakrishnan, L.: Smart characterization of vehicle impact and accident reporting system. In: 2021 7th International Conference Advanced Computing and Communication Systems, ICACCS 2021, pp. 964–968 (2021). https://doi.org/10.1109/ICACCS51430.2021.9441797
3. INEC: Estadisticas de Transporte (2023). https://app.powerbi.com/view?r=eyJrIjoiN2FhOTI xNTktNDA1My00OTQ2LWI2YTltYTIyY2E3NjFiMGY0IiwidCI6ImYxNThhMmU4LW NhZWMtNDQwNi1iMGFiLWY1ZTI1OWJkYTExMiJ9
4. Barona López, G., Velasteguí, L.E.: Automation of industrial processes through Industry 4.0. AlfaPublicaciones **3**(3.1), 84–101 (2021). https://doi.org/10.33262/ap.v3i3.1.80
5. Ammon, D., Stiller, C.: Automated driving. At-Automatisierungstechnik **63**(3), 153–154 (2015). https://doi.org/10.1515/auto-2015-0008
6. Escobar, P., Gonzalez, E., Miño, E., Morales, F.: Object detection in a virtual simulation environment with automated driving toolbox. In: Iberian Conference on Information Systems and Technologies, CISTI, pp. 23–26, June 2021. https://doi.org/10.23919/CISTI52073.2021. 9476414
7. Giurgica, G., Florescu, R.D.: A case study for modeling autonomous driving systems. In: 2020 24th International Conference on System Theory, Control and Computing, ICSTCC 2020 - Proceedings, pp. 745–750 (2020). https://doi.org/10.1109/ICSTCC50638.2020.9259650

Efficient Implementation of a GNSS Base Station with RTK Receiver for Topography

Walter Tana[1], Edgar González[1(✉)] ⓘ, Manuel Montaño[2] ⓘ, and Oscar Gómez[2] ⓘ

[1] Universidad Tecnológica Israel, E4-142, Marieta De Veintimilla y Fco. Pizarro, Quito, Ecuador
{e1725392417,eegonzalez}@uisrael.edu.ec
[2] Universidad Estatal Península de Santa Elena, La Libertad, Ecuador
{mmontano,oscargomez}@upse.edu.ec

Abstract. Recently, low-cost dual-frequency receivers have been evaluated in various positioning applications due to their high-precision capabilities at an affordable cost, positioning them as viable alternatives to high-end geodetic GNSS devices. This research focused on the implementation of a GNSS base station with a Real-time kinematic (RTK) receiver for topographical applications. RTK differential corrections were configured using the u-blox ZED-F9P module, and tests demonstrated their effectiveness in providing accurate measurements in diverse environments. To validate the results, a commercial module was used, and the comparison showed insignificant errors in measurements. Therefore, this study promotes the adoption of low-cost dual-frequency GNSS receivers as viable solutions for high-precision topographical applications.

Keywords: Low-cost receivers · GNSS · Monitoring · RTK · Topography

1 Introduction

The Global Navigation Satellite System (GNSS) has transformed a wide range of applications by enhancing the precision of location measurements from the meter to centimeter level. The use of carrier phase measurements has been essential in this advancement, surpassing the accuracy offered by code-based measurements. Various GNSS positioning techniques exist, including Single Point Positioning (SPP/SPS), Precise Point Positioning (PPP), Differential GNSS (DGNSS), and Real-Time Kinematic (RTK) [1]. SPP/SPS allows for meter-level accuracy through a single independent GNSS receiver, while PPP substantially improves accuracy through post-processing data from independent receivers and other sources. On the other hand, both DGNSS and RTK are differential positioning techniques involving multiple receivers, with one acting as a Base and one or more as Rovers. Rovers use correction data transmitted from the Base to achieve an instant and accurate position solution. Communication in this context can occur through radiofrequency (RF) links with limited range and high costs or over the Internet for transporting data in RTCM format. The Internet-based RTCM Transport Protocol (NTRIP) is widely used to provide real-time RTK services, offering advantages in terms of cost, extended range, and ease of use [2–4].

Á. Rocha et al. (Eds.): WorldCIST 2024, LNNS 985, pp. 82–91, 2024.
https://doi.org/10.1007/978-3-031-60215-3_9

For many applications, high-quality geodetic GNSS instruments have been costly. However, an alternative solution has emerged in the form of low-cost dual-frequency GNSS receivers. These receivers can track more GNSS signals and constellations due to system development, making them more competitive against high-cost geodetic receivers [15]. The combination of these receivers with calibrated ring antennas has further enhanced their competitiveness compared to more expensive geodetic receivers [5]. Researchers have evaluated the performance of these low-cost dual-frequency GNSS receivers in various positioning modes, including relative kinematic positioning, absolute positioning, and real-time kinematic (RTK), in both static and dynamic tests, and have obtained satisfactory results [6–11]. Furthermore, their performance has been evaluated in combination with patch antennas in urban environments, considering static, relative, and absolute positioning modes [12, 13].

The dual-frequency GNSS chip u-blox ZED-F9P and the patch antenna ANN-MB-00 underwent testing in urban and clear-sky conditions for topographical purposes. Under clear-sky conditions and short baselines, centimeter-level accuracy was achieved in relative positioning mode. However, in urban areas, the accuracy reduced to 20 mm horizontally and 15 mm vertically. In RTK mode, the specified manufacturer's precision was not reached as uncertainties increased due to phase center offset (PCO) and phase center variation (PCV) displacements in the patch antennas used [13].

In a similar study, [12] employed the same equipment to conduct research in urban areas using both relative and absolute positioning modes. The results indicate that both methods are suitable for surveys in urban areas under ideal conditions. The static relative method stood out for its higher accuracy and was recommended for applications requiring centimeter-level precision. On the other hand, the PPP method was suggested only in clear-sky environments with few obstacles for satellite signals. In [14], the RTK of the u-blox C099-F9P application board was tested in open sky areas, partially open sky, forests, and urban environments. The researchers concluded that the cost-effective receiver works optimally in open-sky areas but significantly benefits from a geodetic antenna to enhance its positioning performance.

In many developed countries, government entities have established networks of RTK base stations, offering free or affordable RTK correction data streams to the general public. However, in Ecuador, the availability of such networks is limited. In this context, this project focuses on the creation and configuration of GNSS base stations with their own RTK receivers. As technology has advanced, the costs of hardware, software, and internet connectivity have become more accessible, making this effort viable and relevant today.

The article follows the following structure in Sect. 2, the materials and methods are detailed, addressing the description of the devices to be used, their configuration, and programming. In Sect. 3, the results and discussion are presented, including the tests conducted and a comparative analysis. Finally, in Sect. 4, the conclusions of the article are outlined.

2   Materials and Methods

2.1   GNSS Devices

The proposal employs a series of essential devices that play a crucial role in its operation. The ZED-F9P module, serving as the central core of the station, is a dual-frequency GNSS receiver that performs multi-band processing on the same chip. This feature enables centimeter-level geospatial accuracy and real-time resolution of errors such as ambiguity and ionospheric interference. Additionally, a Bluetooth HC-05 module has been integrated to establish wireless communication with mobile devices, adding an extra layer of versatility to the project.

Similarly, the project benefits from the long-distance communication capability provided by the Long Range Radio Modem RFD 915 module, operating in the 915 MHz frequency band. The u-blox multi-band GNSS antenna ensures high precision by receiving signals from multiple satellite constellations, including GPS, GLONASS, Galileo, and BeiDou. For device power and energy management, components such as the Power-Boost 5 V module, the lithium battery charger module TP4056, and a 3.7 V, 2500 mAh LiPo battery are utilized. This set of devices (see Fig. 1) integrates effectively to provide a highly accurate GNSS base station. Together, these components create a comprehensive system capable of conducting precise geospatial measurements and maintaining essential connectivity in the field of topography.

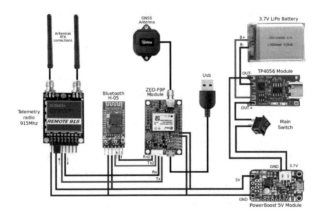

Fig. 1. Base Station Diagram.

2.2   Programming and Configuration of Components

The precise programming and configuration of the components involved in this research are essential to ensure efficient communication among them, enabling the transmission and reception of data necessary for real-time positioning corrections. Additionally, proper configuration ensures synchronization and optimal operation of the system as a whole. To set up a new base station or rover, the process begins with an existing base station. In this particular project, a temporary base station was chosen as the most suitable option. Below, we will detail the basic configurations made with the ZED-F9P Module, for which the following elements are needed:

– A ZED-F9P receiver connected to the computer via a USB cable.
– A location with a clear view of the sky, outdoors, away from obstructions, as accuracy depends on the visibility of as many satellites as possible.
– A high-precision u-blox GNSS multiband L1/L2 antenna.
– An appropriate tripod with a base designed to secure the antennas.
– An SMA extension cable to connect the antenna to the module.
– The "u-center GNSS evaluation software for Windows," downloaded from the official u-blox website.

The first step is to connect the multiband GNSS antenna to the ZED-F9P Module using male and female SMA connectors. Next, the module is connected to the computer via the USB cable, allowing the u-center software to establish a connection with the corresponding COM port. In case of connection issues, it is important to check the port in the Windows Device Manager. A crucial aspect is to ensure that both base receivers and rovers have the latest firmware version from u-blox, such as v1.30. U-blox provides a straightforward update tool, and firmware files are available on their website. It is important to follow the update instructions, unchecking the 'Enter safeboot before update' option to ensure that the module is updated correctly.

Next, the base station is configured. Through u-center, access to the message configuration is obtained, and the necessary RTCM messages for communication with the rover via a LoRa radio link module are enabled. These messages include RTCM3.3 1005, 1074, 1084, 1094, 1124, and 1230 (the latter is enabled to transmit every second). The UART2 baud rate is set to 57600 bps, which is the required speed for LoRa radios.

To monitor the progress and configuration of the base, the SVIN message in the NAV section is enabled. This message provides information about the logging status every second, which is completed when the average standard deviation is less than 5 m. The PVT message can also be checked to ensure that a base has been successfully set up. The base configuration is saved in the module's flash memory. With these settings, the ZED-F9P Module is properly prepared for use as a base station in the real-time positioning system, enabling effective communication with rovers and high precision in geospatial positioning.

The RFD 915 radio modem operates in the 915 MHz ISM band and provides high-power, long-range communication, with ranges of up to 15 km or more.

It offers interfaces for antennas, serial communication, power, and general I/O. Configuration is done through software, such as 'RF Tools-V2.61,' and status LEDs indicate radio activity. For connection, a USB to TTL adapter is used, and baud rates configured in both the ZED-F9P module and the RFD 915 modem must be taken into account. The configuration of the HC-05 module was carried out using a TTL to USB adapter and two HC-05 Bluetooth modules. Through AT commands, the settings of both the slave and master were customized, including baud rate and pairing between devices. Additionally, other AT commands were explored to modify the name, and password, and make specific adjustments, ensuring that the password matches on both devices for successful pairing.

Fig. 2. Base Station with GNSS Antennas and RTK Receiver.

In Fig. 2, the GNSS base station can be observed with two SMA cables extending towards the exterior radio antennas. Once configured as a static base station, it begins to track satellites in its position and calculate corrections for GNSS signals, including adjustments for disturbances in the ionosphere and troposphere, among other sources of error. These correction values are encoded in RTCM format and transmitted to enhance the accuracy of locations reported by GNSS satellites. This process is crucial to maintaining a consistent correction in changing and dynamic environments.

3 Results and Discussion

To test the GNSS, the process begins by installing and securing the base station on a tripod, ensuring it is vertically leveled and securely anchored to the ground. Subsequently, the proper execution of all internal connections is verified, including the addition of an external switch to activate the internal modules. To visualize correction data and establish reference points, the SW Maps application is employed, facilitating the visualization and utilization of positional information from the ZED-F9P chip. For the proper functioning of the application, it is necessary to deactivate the mobile device's location and configure the application as the test location app in the developer options, allowing linkage with the base station.

Connecting the GNSS base station to SW Maps is as simple as searching for and pairing with the device named 'gnssBase' on the mobile device (see Fig. 3). Upon completing the pairing, a request to enter the access key will be prompted. Once the connection is established, the base station integrates as a serial device in SW Maps, enabling the visualization and logging of position data, RTK status, along with the capability to record waypoints, tracks, topographic information, and notes. This functionality proves to be an ideal tool for infrastructure projects and property management.

To conduct the tests, both the rover and the base are powered on. A key visual indicator of proper functioning is the transition of a green LED on the radio modems

Fig. 3. Pairing the Base Station with the Mobile Application.

from a flashing state to a solid one, signaling that the modems have detected each other and are serially transmitting data. Another important indicator is the built-in RTK LED on the ZED-F9P module, functioning as follows: when on; it represents normal satellite mode; if flashing; it indicates an RTK Float; with an accuracy of approximately less than 500 mm; and if off; it represents an RTK Fix; with an accuracy of about 14 mm. In this way, the module is ready for use in the field, as once the rover has acquired RTK corrections, the ZED-F9P sends high-precision coordinates via UART1, UART2, and USB, which are transmitted via radio modem and Bluetooth. Figure 4 illustrates how this is visualized in SW Maps.

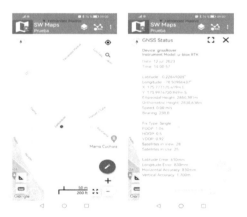

Fig. 4. GNSS Station in Operation.

3.1 System Tests

Testing a precise positioning system and its components is essential to ensure optimal performance. In this context, various comprehensive evaluations were conducted. Firstly, signal tests were performed to measure reception quality and satellite visibility, crucial for determining the system's effectiveness. Subsequently, convergence time tests were carried out to assess how long the system took to achieve an accurate RTK solution. Communication tests were also conducted to verify the integrity of connections between devices. To assess system accuracy, comparison tests were conducted with another set of equipment. Results obtained from static precision tests, along with differentially corrected RTK data, provide a comprehensive view of the system's accuracy and reliability. Finally, a detailed analysis of the results is conducted to determine the overall performance of the system and its suitability for specific applications.

The signal test was conducted to verify the RTK receiver's ability to acquire signals from GNSS satellites, including systems like GPS, GLONASS, Galileo, and BeiDou, resulting in a successful connection. The convergence time test (Table 1) was carried out to evaluate the performance of the RTK receiver in achieving an RTK fix solution upon power-up or after a signal interruption. In this test, it was verified that the time required to achieve convergence remained within acceptable limits as per the specifications required for the project.

Table 1. Convergence time tests.

Tests	Power-up Time (s)	Signal Interruption Time (s)	Convergence Time(s)
Tests 1	11	25	19
Tests 2	10	34	12
Tests 3	8	17	31

With the purpose of verifying the correct transmission of RTCM messages through the radio connection, a test was conducted in which the JST cable from the radio on the rover was disconnected, and a connection was established using a micro USB cable with the 915 MHz radio, creating a COM port. Subsequently, u-center was opened, and the connection with the radio was configured at a transmission speed of 57600 bps, treating it as if it were a GNSS receiver. The packet console was then accessed to monitor the reception and transmission of RTCM messages.

During the evaluation, comprehensive operational tests of the base station were conducted, comparing it with another Emlid REACH RS+ GNSS receiver. This device, operating similarly to the one used in this project, was positioned at the same point previously occupied by the base station. The Emlid REACH RS+ is known for being a high-precision GNSS receiver that uses LoRa technology to provide reliable connections with a baseline range of up to 8 km and a transmission power of up to 100 mW. Additionally, it features an internal global quad-band LTE modem and integrated Wi-Fi connectivity, along with 16 GB of internal storage for data. It also offers compatibility with RTCM 3.x and NMEA 0183 standards for data input and output. These qualities make it an ideal choice for conducting meaningful comparisons.

The results of the static precision test involved establishing a fixed static control point using the Emlid REACH RS+ device, whose coordinates were compared with those obtained by the GNSS base station. The main objective was to assess the system's accuracy in terms of both horizontal and vertical errors, expressed in meters for a clearer and more effective interpretation. In this process, four main measurement points were selected, where the average error in the measurement is approximately 0.009 m. The differential correction test was conducted using the GNSS base station with previously established high-precision coordinates. The purpose of this test was to take real-time measurements while moving with the RTK receiver, recording coordinates at various points of interest. The aim of this process was to compare the obtained coordinates with the already known reference coordinates, thus allowing an assessment of accuracy in motion, resulting in an average error of 0.015.

3.2 Results Analysis

The analysis of the results obtained at each evaluation stage is crucial to determine the performance and suitability of the implemented GNSS system. Firstly, in the signal test, it was confirmed that the GNSS receiver successfully acquired signals from GNSS satellite systems, including GPS, GLONASS, Galileo, and BeiDou. This is crucial to ensure accurate position determination and navigation, and the results indicated high-quality signal reception from these systems in both desktop and field test environments.

The time-to-convergence test revealed that the RTK receiver achieves rapid convergence, which is essential for applications requiring early acquisition of position accuracy. This feature is particularly valuable in situations where an efficient system response is needed. The results indicated that the RTK receiver can achieve the required accuracy in an appropriate time, validating its utility in real-time applications.

Regarding the communication evaluation, the system's ability to transmit and receive RTCM messages through the radio connection was successfully demonstrated. The RTK receiver was able to transition from RTK Float to RTK Fix status after receiving only a few seconds of RTCM data through the radio connection. This confirms effective communication between internal components and the system's ability to provide high-precision data continuously.

Finally, the accuracy results were divided into two parts: static positioning accuracy (see Table 2) and differential corrections. In static positioning accuracy, small differences were observed in the coordinates obtained compared to the reference coordinates at each measurement point. These small deviations indicate a shift in the recorded position relative to the reference position, but these discrepancies are relatively minimal.

In the evaluation of differential corrections, the coordinates obtained with the RTK receiver were compared with the reference coordinates established by the GNSS base station. The differences between these coordinates were on the order of millimeters, indicating high precision of the RTK receiver in obtaining position in motion. These differences could be attributed to measurement errors and environmental conditions, but overall, the results demonstrate good performance of the RTK receiver for real-time precision applications.

Table 2. Results of static accuracy.

Measurement Point	Emlid REACH RS2+	GNSS Base Station	Difference (m)
Tests 1	−78,44628822/	−78.44627623/	Lat: 0.012
	−0.00429800	−0.00428802	Lon: 0.010
Tests 2	−78,44621965/	−78.44620566/	Lat:: 0.014
	−0,00411033	−0.00410132	Lon: 0.009
Tests 3	−78,44726067/	−78.44725567/	Lat::-0.005
	−0,00386050	−0.00386650	Lon: 0.006

4 Conclusions

This research has delved into the fundamental principles of GNSS, focusing on systems such as GPS, GLONASS, BeiDou, and Galileo, providing a solid understanding of global positioning technology. The successful implementation of the u-blox ZED-F9P RTK receiver, coupled with real-time (RTK) differential corrections setup, has paved the way for precise real-time position measurements. The programming, configuration, and effective communication process among electronic components, including the RTK receiver, GNSS chip, radio modem, and Bluetooth modules, have resulted in accurate real-time position measurements, laying the foundation for future topography projects requiring precision and wireless communication.

References

1. Teunissen, P.J.G., Montenbruck, O.: Springer Handbook of Global Navigation Satellite Systems, vol. 10. Springer (2017)
2. Mahato, S., Santra, A., Dan, S., Banerjee, P., Kundu, S., Bose, A.: Point positioning capability of compact, low-cost GNSS modules: a case study. IETE J. Res. **69**(7), 4099–4112 (2023)
3. Xiuqiang, Z., Xiumei, Z., Yan, C.: Implementation of carrier phase measurements in GPS software receivers. In: 2013 International Conference on Computational Problem-solving (ICCP), pp. 338–341 (2013)
4. Shamaei, K., Kassas, Z.M.: Sub-meter accurate UAV navigation and cycle slip detection with LTE carrier phase measurements. In: Proceedings of the 32nd International Technical Meeting of the Satellite Division of the Institute of Navigation (ION GNSS+ 2019), pp. 2469–2479 (2019)
5. Ardusimple GNSS Antennas. https://www.ardusimple.com/. Accessed 22 Oct 2023
6. Lăpădat, A.M., Tiberius, C.C.J.M., Teunissen, P.J.G.: Experimental evaluation of smartphone accelerometer and low-cost dual frequency GNSS sensors for deformation monitoring. Sensors **21**(23), 7946 (2021)
7. Tunini, L., Zuliani, D., Magrin, A.: Applicability of cost-effective GNSS sensors for crustal deformation studies. Sensors **22**(1), 350 (2022)
8. Romero-Andrade, R., Trejo-Soto, M.E., Vázquez-Ontiveros, J.R., Hernández-Andrade, D., Cabanillas-Zavala, J.L.: Sampling rate impact on precise point positioning with a Low-Cost GNSS receiver. Appl. Sci. **11**(16), 7669 (2021)

9. Hamza, V., Stopar, B., Ambrožič, T., Sterle, O.: Performance evaluation of low-cost multi-frequency GNSS receivers and antennas for displacement detection. Appl. Sci. **11**(14), 6666 (2021)

10. Krietemeyer, A., van der Marel, H., van de Giesen, N., ten Veldhuis, M.-C.: High quality zenith tropospheric delay estimation using a low-cost dual-frequency receiver and relative antenna calibration. Remote Sens. **12**(9), 1393 (2020)

11. Semler, Q., Mangin, L., Moussaoui, A., Semin, E.: Development of a low-cost centimetric GNSS positioning solution for Android applications. Int. Arch. Photogramm. Remote Sens. Spat. Inf. Sci. **42**, 309–314 (2019)

12. Romero-Andrade, R., Trejo-Soto, M.E., Vega-Ayala, A., Hernández-Andrade, D., Vázquez-Ontiveros, J.R., Sharma, G.: Positioning evaluation of single and dual-frequency low-cost GNSS receivers signals using PPP and static relative methods in urban areas. Appl. Sci. **11**(22), 10642 (2021)

13. Wielgocka, N., Hadas, T., Kaczmarek, A., Marut, G.: Feasibility of using low-cost dual-frequency GNSS receivers for land surveying. Sensors **21**(6), 1956 (2021)

14. Janos, D., Kuras, P.: Evaluation of low-cost GNSS receiver under demanding conditions in RTK network mode. Sensors **21**(16), 5552 (2021)

15. Rakhmanov, A., Wiseman, Y.: Compression of GNSS data with the aim of speeding up communication to autonomous vehicles. Remote Sens. **15**(8), 2165 (2023)

Refining Cyber Situation Awareness with Honeypots in Case of a Ransomware Attack

Jouni Ihanus[1]([✉])[ID], Tero Kokkonen[2][ID], and Timo Hämäläinen[1][ID]

[1] Faculty of Information Technology, University of Jyväskylä, Jyväskylä, Finland
jouni.e.i.ihanus@student.jyu.fi, timo.hamalainen@jyu.fi
[2] Institute of Information Technology, JAMK University of Applied Sciences,
Jyväskylä, Finland
tero.kokkonen@jamk.fi

Abstract. The cyber threat landscape is vast and unstable. One of the top threats in the present moment is ransomware, which is constantly spreading in prevalence. To protect organisations' cyber operating environment, ability to perceive elements relating to this threat is crucial. At the same time, many security controls face challenges in terms of fidelity of the security events. In this paper, honeypot technology is studied to support situation awareness in case of a ransomware attack. Especially detection capabilities of the honeypots are considered from the perspective of technical characteristic of ransomware. As a conclusion, we propose a construction model for enhancing cyber situation awareness using honeypots during various stages of a ransomware attack. Additionally, the analysed results are explained with identified future research topics.

Keywords: Honeypots · Cyber Situation Awareness · Ransomware

1 Introduction

One of the top threats organisations face in the present moment is ransomware, which is constantly spreading in prevalence. [23] defines ransomware as "malware which encrypts or manipulates files located in digital device and demands a ransom from the user for decryption". Ransomware can also threaten to disseminate or disclose confidential information [23]. This kind of malware poses high operational and financial risk for organisations, which has been realised in several high-level cyber-attacks [28,32,34]. At the same time multiple threat reports show a steady increase in ransomware attacks. This phenomenon can be observed in terms of different ransomware variants and popularity of the attack type [11,13,27]. According to the Cyber Security Centre Finland, the number of ransomware attacks has increased approximately 30% yearly [20], also [5] indicates rise and progress of sophistication of ransom attacks. Therefore, it is crucial to **perceive the elements** related to the ransomware attack, which is considered the initial step in achieving situation awareness in any dynamic system, as stated by [10]. This observation highlights the importance of the research field.

Á. Rocha et al. (Eds.): WorldCIST 2024, LNNS 985, pp. 92–101, 2024.
https://doi.org/10.1007/978-3-031-60215-3_10

Modern strong encryption techniques are used in ransomware implementations. Especially open-source implementations of encryption tools are easy to achieve and accommodate as part of the ransomware [26]. Ransomware is also a major threat to the critical infrastructure; hence, ransomware detection and protection shall be considered as part of the critical infrastructure protection [18]. Ransomware is not limited to infecting traditional computer networks; authors of the paper [17] introduce study of ransomware on connected vehicles. Detection of ransomware before it starts to encrypt data on the infected system is a complex task, authors of [25] use blockchain technology to detect ransomware by a record of the pre-encryption and post-encryption behaviour. Blockchain technology is used to detect and defend against ransomware in critical infrastructure of smart healthcare [31].

In addition, the modern honeypot technologies are used for detecting ransomware attacks. Study [19] applies honeypot folder for detecting ransomware by monitoring changes, while [30] used honeypots for identifying ransomware activities on a website. Honeypots can also be used for the detection of ransomware in several infrastructures: the study [7] proposes honeypot based on MikroTik devices. Authors of paper [24] introduce Intrusion Detection Honeypot (IDH) for detecting ransomware in IoT infrastructure. The study [35] analysed several ransomware attacks and created a file-based honeypot method for identifying ransomware processes, while study [21] used honeypot technology for detecting ransomware with useful data of identified ransomware type.

In our earlier papers [15,16], authors presented how medical devices can be modelled with honeypots by the created conceptual framework relating to detection capabilities of the honeypots. This research expands upon the previous studies by examining the honeypot's ability to detect ransomware attacks. It is important to note that the scope of this research extends beyond medical devices. The main contributions of this paper are: how to enhance the cyber situation awareness with honeypot technology in different stages of a ransomware attack.

The rest of the paper is organised as follows: First, the concept of ransomware is described in Sect. 2. It is followed by Sect. 3 which illustrates the capabilities of the honeypot technology in ransomware detection. Section 4 describes the modelled construction for the study. The main results can be found in the Sect. 5, and the whole study is concluded with identified future work in the Sect. 6.

2   Ransomware

To analyse the capabilities of honeypot technology in the event of a ransomware attack, one must comprehend the stages typically involved in such attacks. This chapter aims to provide this fundamental knowledge, utilising conventional frameworks to ensure clarity and comprehension. The aim is to answer the question: How are ransomware attacks typically structured?

2.1   Taxonomy of a Ransomware Attack

Taxonomy of the cyber attack can be modelled based on multiple frameworks. For example, study of [9] divides ransomware attacks based on Lockheed Martin's Cyber Kill Chain (CKC) while [1] and [6] seem to follow selected parts of the Mitre ATT&CK Matrix. Both of these frameworks are highly well-known. In this study the authors will follow the attack stages presented in [6], which is more focused on the attack type while partly correlating with CKC and ATT&CK frameworks. However, it is worth noting that not all ransomware attacks or other types of attacks adhere to all the stages outlined in modelling frameworks. Based on a selected framework, a ransomware attack can be divided into seven different stages, as follows:

Stage 1. Reconnaissance stage includes actions to collect relevant information about the target. This information creates the base for following stages, for example to select the best tool to successfully penetrate the target environment [9,14]. In addition to the technical review, this stage can include an economic evaluation of the victim. This information can aid in assessing operational weaknesses and determining the appropriate ransom amount [1,6].

Stage 2. Initial access utilises the goal to gain access to the target network for further activity. Access can be by any means, including, but not limited to, the following [1,6,8]: Phishing, credential stuffing/reuse, vulnerability exploitation, insider threat, social engineering, infected websites, and flash drives.

Stage 3. Escalation and lateral movement stage is said to be the most complicated part of the ransomware attack [1]. This stage deepens the adversary's understanding of target organisation's network. Typical actions are network and vulnerability mapping and exploitation of the related systems. This increases the level of control and access that an attacker possesses. The process brings the attacker closer to their ultimate goal of accessing the data they seek to steal or use for blackmail [1,6].

Stage 4. Data collection and exfiltration are critical components of so called secondary extortion. In this approach, the victim is initially blackmailed with encrypted data and subsequently threatened to disclose any sensitive information obtained during this stage. The attacker is essentially seeking all sensitive data vital to the victim, such as financial, accounting, client, and intellectual property information. This information is then gathered and sent to a server controlled by the attacker [1,6].

Stage 5. Degradation of recovery and system security services stage is motivated by the desire to remain undetected for as long as possible within the victim's network for further research purposes. The aim is to impact the active security solutions in the target environment in order to sustain a foothold. Additionally, by compromising the victim's backup systems extortion cannot be ignored, as the original data cannot be restored [6].

Stage 6. Deployment, execution, and encryption is implemented when the attacker has extracted the relevant data from the victim's network. In Cyber Kill-chain based models this stage is called *Actions on objectives* [9,14]. Practical measures include the encryption of relevant data and the delivery of a ransom demand to the affected party [6].

Stage 7. Recovery and retrospective stage emphasises the importance of implementing comprehensive recovery protocols from the victim's perspective. It is crucial to comprehend the techniques and weaknesses employed by the attacker. Failing to do so could result in the organisation being confronted with persistent actor and possibility of re-encryption and exfiltration of sensitive data [6].

3 Capabilities of Honeypot Technology in Ransomware Detection

To further analyse the positioning of honeypot technology among different technologies, the ransomware research taxonomy can be used. According to [22], most common area of research in recent studies focuses on **detection of ransomware attacks**. Other areas covered by the study included the classification, mitigation, prediction, and prevention of the attack. Although these areas partly overlap, detection appears to have generated the most controversy among researchers. Detection can be defined as a process to identify the ransomware on a system or network, which can be achieved with multiple methods. One method employed in these studies is known as **decoy-based technique**, comprising techniques associated with honeypots. The scope of honeypot technologies includes decoy systems (honeypots) and decoy files (honeyfiles). Both implementations can be utilised to identify the threats described in this study. To briefly explore the differences between these two honey-based methods, it can be concluded that honeypots usually imitate a valuable network resource or system [22]. Conversely, honeyfiles utilise a system for monitoring essential file and alarm functions [33]. In this study, the term honeypot is utilised to refer to both methods, when it is not pertinent to the context.

Previous research and literature have identified features that apply to decoy-based technology [1,4,22]. These features help to assess monitoring objectives for honeypot usage regarding ransomware attacks. In this study, we combine these features with attack stages to evaluate honeypot technology's suitability for detecting ransomware. The list of the identified features include: file access, file system activity, user data, system access, command detection, binary and tool collection, understanding the methodology, and distraction.

The above-presented features include multiple system-level elements that are accessible through the decoy system. It should be also noted that the identified features include observation of attack methodology, which can increase understanding of the context.

4   Construction

In the preceding chapters, a typical taxonomy of ransomware attack was presented, along with the key capabilities of honeypots related to the attack type in question. This chapter aims to combine these elements to construct a method for detecting ransomware attacks using honeypots during various stages of the attack. In terms of a situation awareness, it is crucial to perceive elements related to an attack in order to further process comprehension of the current situation.

The irrecoverable nature of a ransomware attack creates pressure to identify an attack at early stages. For this reason, it is essential to assess the applicability of detection technology throughout various stages of an attack.

Stage 1. Reconnaissance

To identify potential targets for an organisation's internet-facing system on a technical level, an adversary should utilise active methods such as port scanning. As honeypots can decoy multiple services, these techniques are notably pertinent to the scope of this study. Extensively logging active connections is crucial to detecting pertinent information for stage one, allowing for the identification of low-level reconnaissance.

In terms of visibility, honeypots can be located on multiple levels of network to detect these actions. It should be noted that exposure to a direct internet segment may increase the number of events detected by the honeypot. The root cause of this issue is the well-known ubiquitous internet background noise. This can make evaluating the context and relevance of the events difficult. However, by using honeypots in this manner, one can analyse attack techniques to develop new defence measures [4]. These externally facing honeypots can, for example, reveal the type of exploits attackers are currently using [1]. Understanding these methods can be used to implement more targeted honeypots for stage two. The use of honeypots can also divert the attention of the adversary, thus reducing their focus on the actual assets [4].

To summarise the practicality of honeypots in this stage, Sect. 3 outlines several detection features which can be utilised. Nevertheless, it is vital to acknowledge the potential for resource exhaustion when exposed to the direct internet segment.

Stage 2. Initial Access

On the network level, system access can be executed through multiple protocols. Honeypots can help with many of them [12]. However, the key challenge in this stage is the selection of the initial access method to be decoyed. The methods presented in Sect. 2.1 include elements such as vulnerability exploitation and infected websites can be used if they are relevant for one's technical environment. One can also expose some of the unfortunately generic internet-facing services. One example presented by [1] is the implementation of internet-facing RDP in an isolated network segment for this purpose. Actions against these decoy methods can be seen more finer-grained than typical port-scanning, and therefore do not raise as many events. Reducing the logging level of these actions is acceptable in comparison to stage 1. This lowers the risk for resource exhaustion.

To summarise the practicality of honeypots in this stage, the usability of the detection features provided by the honeypot is strongly related to the choice of decoy method. One should analyse which services are credible in one's environment to lure the adversary into taking action.

Stage 3. Lateral Movement

Inside the organisation's network, the adversary must locate the actual assets they will use for extortion. From a defender's point of view, this is one of the best stages to use decoy-based solutions for detection. In order to reach the sixth stage (deployment), the attacker must remain undetected for a significant period of time, which can provide several opportunities to catch the adversary. However, adversaries are aware that organisations may have honeypots on their network and will be searching for them. Thus, honeypots should not be named too obviously [1]. Another clear obstacle is the process of fingerprinting, whereby an adversary can recognise the decoy according to its characteristics or behaviours [29]. Furthermore, it should be remembered that a decoy-based solution cannot ensure that the adversary will attack the honeypot first [2]. Therefore, decoys must merge with the rest of the environment.

In the internal network, background noise is noticeably absent compared to stage one. There are typically no unknown port-scanning actions taking place within the internal network, thus making all actions against the honeypot suspicious. As a result, false positives are significantly reduced [29]. It is in this environment that honeypots prove to be particularly useful in detecting ransomware attacks. Almost all the typical assets of an organisation and network can be modelled. These may incorporate, but are not limited to, a honey backup server, file server, domain controller, and endpoints [1].

One notable technical manner relating to this stage is that criminals are increasingly relying on the use of "living off the land" (LoL) tools and binaries [1,6]. The use of these native operating system commands is a challenge in terms of detection for many detection solutions. This is just one example of the evasion techniques that attackers implement to reach their ultimate goal of data exfiltration and extortion [9]. By utilising a feature-rich honeypot, one may develop command detection tools in order to identify any abnormal activity. Usually, this is accomplished at the receiving end of honeypot-generated events, such as in SIEM. Naturally, these commands can be monitored in real systems as well, however, the number of false-positive events increases due to their potential use for legitimate tasks.

In summary, to maximise the practicality of honeypots at this stage, one can include any of the features provided by honeypots. As there is no background noise within the organisation's internal network, honeypots can generate reliable events even in large deployments.

Stage 4. Data Collection and Exfiltration

As an additional tool for detecting malicious activity, honeyfiles can be particularly beneficial during the fourth stage. These decoy files serve as notifications to the defender when accessed or modified [33]. They can be placed strategically in fileservers, workstations, or other relevant locations within one's network. It is worth noting that adversaries typically employ keywords to discover significant assets. This factor should be considered while producing honeyfiles [1].

Data can be collected from various locations, making honeyfiles a cost-effective solution with relatively broad detection capabilities. Honeyfiles are easier to distribute than a full system honeypot and therefore simpler to implement. However, this solution lacks the contextual information provided by high-interaction honeypots.

Stage 5. Degradation of Recovery and System Security Services
Tampering with backup systems, the firewall, or other security measures can disrupt recovery and compromise system security services. Such interference may involve the use of system commands intended for operation system level management [6]. Honeypots can be used to alert specified system command as presented in stage three. As a summary, this stage specifically enhances the command detection capability of the honeypots.

Stage 6. Deployment, Execution, and Encryption
As the adversary has extracted valuable assets and data from the organisation's network, they initiate the deployment of ransomware. While this stage may appear dramatic to the victim, the technical activities are relatively standard compared to the previously mentioned stages. However, in this stage an adversary can try to use multiple evasion techniques as presented in [9]. Especially detection features presented in stages three to five are also very useful at this stage.

Stage 7. Recovery and Retrospective
Section 2.1 highlights the significance of stage seven in comprehending the nature of ransomware attacks. However, a major challenge faced is that script based ransomware usually eliminates the original script and encryption-related tools [9], which makes it difficult to investigate the aftermath, owing to the lack of evidence availability. However, honeypots can store attack information outside of the actual file system, allowing for analysis of technical details during the recovery and retrospective stages.

To summarise stage seven, it is noteworthy that information gathered by the honeypots is often exported to a different location. This enables post-attack analysis to take place from a reliable source even if the real decoy system has been breached. The same principle applies to features such as binary and tool collection presented in Sect. 2.1. This feature is essential in ensuring that the adversary is no longer persistent within organisational systems and presents no longer a threat with re-encryption.

5   Results

To gain a comprehensive understanding of the issues that ransomware attacks can create for organisations, it is necessary to differentiate between *network-centric* and *domain-centric* cyber situational awareness. This involves separating technological data and situational awareness information from each other, as demonstrated in [3]. While technological data may contain typical indications of compromise (IoCs) such as IP addresses and file hashes, situational awareness information encompasses contextual details such as motivation and methods of attack. Decoy-based detection methods employing honeypots and honeyfiles are vital for improving situational awareness in ransomware attacks. These methods not only provide high-quality technical data but also valuable contextual information to support analysis. The construction model presented in this study provides insight into the capabilities of honeypots at different stages of a ransomware attack. Thus, this methodology provides essential contextual details regarding the attack's characteristics and the actual actions taken by the adversary.

6   Conclusion

It can be argued that ransomware attacks are technically quite similar to any other malware attack. It is the blackmail aspect and the motivation behind it that remains static in case of a ransomware attack. It is important to understand that *motivation* determines the attacker's approach and actions, making this type of attack unique. While techniques change, the motivation remains the same, giving the defender a static element to rely on. Therefore, detecting and mitigating ransomware attacks can benefit from a detection mechanism that can respond to the motivational aspect and dynamically detect different technological attack approaches. Honeypots can be tailored to meet this motivational need and detect relevant technical context data from various types of attack. It can be concluded that according to the presented construction model, honeypots can improve situation awareness in case of a ransomware attack.

For further research, it is crucial to examine the drawbacks of decoy-based solutions. A notable weakness is their limited field of vision; they can only detect malicious activities directed towards them. Another significant challenge involves fingerprinting, as it can result in the honeypot being detected, thereby compromising its effectiveness.

Acknowledgement. This research was partially funded by the Resilience of Modern Value Chains in a Sustainable Energy System project, co-funded by the European Union and the Regional Council of Central Finland (grant number J10052).

The authors would like to thank Ms. Tuula Kotikoski for proofreading the manuscript.

References

1. Liska, A.: Ransomware: Understand. Prevent. Recover, Future US LLC (2023)
2. Alqahtani, A., Sheldon, F.T.: A survey of crypto ransomware attack detection methodologies: An evolving outlook. Sensors (2022). https://doi.org/10.3390/s22051837
3. Bass, T.: Intrusion detection systems and multisensor data fusion. Commun. ACM **43**(4), 99–105 (2000). https://doi.org/10.1145/332051.332079
4. Bin Sulaiman, R., Rahi, M.: A detailed study on web-based-honeypot to propose mitigation framework in web application. SSRN Electron. J. (2019). https://doi.org/10.2139/ssrn.3723098

5. Brooks, C.: Alarming Cyber Statistics for Mid-Year 2022 That You Need To Know (2022). https://www.forbes.com/sites/chuckbrooks/2022/06/03/alarming-cyber-statistics-for-mid-year-2022-that-you-need-to-know/?sh=4a6b79687864

6. Center, T.A.R.: Decoding the dna of ransomware attacks: unveiling the anatomy behind the threat. https://www.trellix.com/about/newsroom/stories/research/decoding-the-dna-of-ransomware-attacks/. Accessed: 20 Oct 2023

7. Ceron, J.M., Scholten, C., Pras, A., Santanna, J.: Mikrotik devices landscape, realistic honeypots, and automated attack classification. In: NOMS 2020 - IEEE/IFIP Network Operations and Management Symposium, pp. 1–9 (2020). https://doi.org/10.1109/NOMS47738.2020.9110336

8. Chayal, N., Saxena, A., Khan, R.: A review on spreading and forensics analysis of windows-based ransomware. Ann. Data Sci. (2022). https://doi.org/10.1007/s40745-022-00417-5

9. Dargahi, T., Dehghantanha, A., Nikkhah Bahrami, P., Conti, M., Bianchi, G., Benedetto, L.: A cyber-kill-chain based taxonomy of crypto-ransomware features. Journal of Computer Virology and Hacking Techniques **15** (2019). https://doi.org/10.1007/s11416-019-00338-7

10. Endsley, M.: Toward a theory of situation awareness in dynamic systems. Hum. Factors **37**(1), 32–64 (1995). https://doi.org/10.1518/001872095779049543

11. Fortinet Inc.: Global Threat Landscape Report A Semiannual Report by Forti-Guard Labs - August 2022. Tech. rep. Accessed 2 Oct 2022

12. Franco, J., Aris, A., Canberk, B., Uluagac, A.S.: A survey of honeypots and honeynets for internet of things, industrial internet of things, and cyber-physical systems. IEEE Communications Surveys & Tutorials (2021). https://doi.org/10.1109/COMST.2021.3106669

13. Humayun, M., Jhanjhi, N., Alsayat, A., Ponnusamy, V.: Internet of things and ransomware: evolution, mitigation and prevention. Egyptian Inform. J. **22**(1), 105–117 (2021). https://doi.org/10.1016/j.eij.2020.05.003

14. Hutchins, E., Cloppert, M., Amin, R.: Intelligence-driven computer network defense informed by analysis of adversary campaigns and intrusion kill chains. Leading Issues Inform. Warfare Secur. Res. **1** (2011)

15. Ihanus, J., Kokkonen, T.: Modelling medical devices with honeypots. In: Internet of Things, Smart Spaces, and Next Generation Networks and Systems, pp. 295–306. Springer, Cham (2020)

16. Ihanus, J., Kokkonen, T., Hämäläinen, T.: Modelling medical devices with honeypots: a conceptual framework. In: Information Systems and Technologies, pp. 145–155. Springer, Cham (2022)

17. Malik, A.W., Anwar, Z., Rahman, A.U.: A novel framework for studying the business impact of ransomware on connected vehicles. IEEE Internet Things J. (2022). https://doi.org/10.1109/JIOT.2022.3209687

18. Mead, N.R.: Critical infrastructure protection and supply chain risk management. In: 2022 IEEE 30th International Requirements Engineering Conference Workshops (REW), pp. 215–218 (2022). https://doi.org/10.1109/REW56159.2022.00047

19. Moore, C.: Detecting ransomware with honeypot techniques. In: 2016 Cybersecurity and Cyberforensics Conference (CCC) (2016). https://doi.org/10.1109/CCC.2016.14

20. NCSC, Finland: Threat level in cyber environment has risen - activity towards Finland has increased (2022). https://www.traficom.fi/en/news/threat-level-cyber-environment-has-risen-activity-towards-finland-has-increased

21. Pascariu, C., Barbu, I.D.: Ransomware honeypot: honeypot solution designed to detect a ransomware infection identify the ransomware family. In: 2019 11th International Conference on Electronics, Computers and Artificial Intelligence, pp. 1–4 (2019). https://doi.org/10.1109/ECAI46879.2019.9042158

22. Razaulla, S., et al.: The age of ransomware: a survey on the evolution, taxonomy, and research directions. IEEE Access **PP**, 1 (2023). https://doi.org/10.1109/ACCESS.2023.3268535

23. Sanastokeskus ry. https://sanastokeskus.fi/tiedostot/pdf/Kyberturvallisuuden_sanasto.pdf?file=pdf/Kyberturvallisuuden_sanasto.pdf. Accessed: 2 Oct 2022

24. Sibi Chakkaravarthy, S., Sangeetha, D., Cruz, M.V., Vaidehi, V., Raman, B.: Design of intrusion detection honeypot using social leopard algorithm to detect iot ransomware attacks. IEEE Access (2020). https://doi.org/10.1109/ACCESS.2020.3023764

25. Singh, A., Ali, M.A., Balamurugan, B., Sharma, V.: Blockchain: tool for controlling ransomware through pre-encryption and post-encryption behavior. In: 2022 Fifth International Conference on Computational Intelligence and Communication Technologies (2022). https://doi.org/10.1109/CCiCT56684.2022.00107

26. Smith, D., Khorsandroo, S., Roy, K.: Machine learning algorithms and frameworks in ransomware detection. IEEE Access (2022). https://doi.org/10.1109/ACCESS.2022.3218779

27. SonicWall, Inc.: 2022 Cyber Threat Report. Technical report. Accessed 27 Sep 2022

28. TechTarget, A.C.: Cloudnordic loses most customer data after ransomware attack. https://www.techtarget.com/searchsecurity/news/366549773/CloudNordic-loses-most-customer-data-after-ransomware-attack. Accessed: 28 Aug 2023

29. Titarmare, N., Hargule, N., Gupta, A.: An overview of honeypot systems. Int. J. Comput. Sci. Eng. **7**, 394–397 (2019). https://doi.org/10.26438/ijcse/v7i2.394397

30. Venkatesh, J., Vetriselvi, V., Parthasarathi, R., Subrahmanya V.R.K., Rao, G.: Identification and isolation of crypto ransomware using honeypot. In: 2018 Fourteenth International Conference on Information Processing (2018). https://doi.org/10.1109/ICINPRO43533.2018.9096875

31. Wazid, M., Das, A.K., Shetty, S.: BSFR-SH: blockchain-enabled security framework against ransomware attacks for smart healthcare. IEEE Trans. Consum. Electron. (2022). https://doi.org/10.1109/TCE.2022.3208795

32. Wired, A.G.: The untold story of notpetya, the most devastating cyberattack in history. https://www.wired.com/story/notpetya-cyberattack-ukraine-russia-code-crashed-the-world/. Accessed 6 Oct 2022

33. Yuill, J., Zappe, M., Denning, D., Feer, F.: Honeyfiles: deceptive files for intrusion detection. In: Proceedings from the Fifth Annual IEEE SMC Information Assurance Workshop (2004). https://doi.org/10.1109/IAW.2004.1437806

34. Zdnet, C.C.: Norsk hydro ransomware incident losses reach $40 million after one week. https://www.zdnet.com/article/norsk-hydro-ransomware-incident-losses-reach-40-million-after-one-week/. Accessed 6 Oct 2022

35. Zhuravchak, D., Ustyianovych, T., Dudykevych, V., Venny, B., Ruda, K.: Ransomware prevention system design based on file symbolic linking honeypots. In: 2021 11th IEEE International Conference on Intelligent Data Acquisition and Advanced Computing Systems: Technology and Applications (2021). https://doi.org/10.1109/IDAACS53288.2021.9660913

Machine Learning for Forensic Occupancy Detection in IoT Environments

Guilherme Dall'Agnol Deconto[1]([envelope]), Avelino Francisco Zorzo[1],
Daniel Bertoglio Dalalana[1], Edson Oliveira Jr[2], and Roben Castagna Lunardi[3]

[1] Pontifical Catholic University of Rio Grande do Sul, Porto Alegre, RS, Brazil
{g.dallagnol,daniel.bertoglio}@edu.pucrs.br, avelino.zorzo@pucrs.br
[2] State University of Maringá, Maringá, PR, Brazil
edson@din.uem.br
[3] Federal Institute of Education, Science and Technology of Rio Grande do Sul,
Porto Alegre, RS, Brazil
roben.lunardi@restinga.ifrs.edu.br

Abstract. The adoption of the Internet of Things (IoT) has brought many advantages, but it also presents challenges for the field of Digital Forensics. The heterogeneity of the data directly affects the investigative process in scenarios involving IoT applications. Through the analysis of a comprehensive and heterogeneous dataset collected from IoT devices, this study analyzes the use of machine learning algorithms to detect specific patterns to estimate the number of people in physical environments involving IoT devices. In this work, we discuss the use of Machine Learning approaches to enhance criminal investigations based on data collected from IoT environments. The experimental evaluation not only showcases the potential enhancement of Digital Forensics through the utilization of IoT data but also serves to emphasize the effectiveness of machine learning-based approaches in these environments.

Keywords: Internet of Things · Digital Forensics · Machine Learning

1 Introduction

The term Internet of Things (IoT) has been increasingly used to refer to a variety of devices and equipment with low computational power that are connected to the Internet. Data collected from sensors and actuators in IoT environments can be used to remotely monitor and control them. Additionally, IoT devices offer significant advantages in various sectors, including industry, healthcare, transportation, and agriculture, enabling process optimization and improving product quality [17].

The intersection between IoT and Digital Forensics presents a complex and challenging dynamic. As the IoT continues to expand and diversify, an increasing number of connected devices generate a massive amount of data, becoming a valuable source of evidence for forensic investigations. However, the heterogeneous nature and large volume of data from these devices require specific and customized approaches. The convergence of Digital Forensics with Machine

Á. Rocha et al. (Eds.): WorldCIST 2024, LNNS 985, pp. 102–114, 2024.
https://doi.org/10.1007/978-3-031-60215-3_11

Learning offers unique opportunities to enhance investigation and forensic analysis in IoT environments, allowing the analysis of patterns, detection of suspicious events, and a deeper understanding of the investigated scenarios [3,25].

Although there are numerous researches focused on the context of IoT Forensics, there is still a lack of tools designed and developed to handle the large volume and heterogeneity of data generated by IoT devices [12]. Furthermore, these tools can allow for an accurate assessment of the consequences of the loss or destruction of a specific IoT device. Therefore, it is necessary to develop specific tools to aid in conducting investigations in this scenario, as current tools are not able to meet this demand [14,15].

Detecting patterns in IoT data can be instrumental in uncovering criminal activities within smart environments. Such data can provide valuable information on the *modus operandi* and potential culprits. This research is part of our effort to identify patterns in physical environments, ultimately contributing to the improvement of crime detection in physical smart environments. Specifically, this study aims to **evaluate and discuss the adoption of machine learning models to accurately estimate the occupancy of a given environment using data collected from IoT devices** in the field of Digital Forensics.

To achieve the objectives outlined in this paper, we used a publicly available IoT multisensor room dataset for comprehensive data analysis. The use of supervised machine learning techniques, including Random Forest, Decision Tree, and XGBoost, forms a crucial aspect of our approach, facilitating accurate detection of room occupancy. Our methodology involves the deployment of these machine learning models, and we conduct comparisons to assess their efficacy in this context. The collected results were meticulously evaluated, and the discussion shed light on the insights gained, providing a deeper understanding of the performance and implications of the techniques used.

2   Related Work

Several researchers emphasize the urgency of developing innovative approaches to address the complex scenario of Digital Forensics in the IoT. The exponential growth in the number of connected devices has generated a substantial volume of data, necessitating specialized strategies for the collection, analysis, and interpretation of this information. The challenge lies not only in the quantity of data but also in the diversity and interdependence of these devices, adapting forensic techniques essential to ensure effectiveness in investigation and the preservation of the integrity of digital evidence. In this dynamic context, the continuous evolution of Digital Forensics practices in IoT is crucial to keep pace with and overcome the emerging challenges in this field [9,10,19,21,24].

Adedayo [1] highlights the complexity associated with one of the primary challenges faced by Digital Forensics: the considerable volume of data that requires analysis. The authors underscore the imperative need to reassess our approaches in the field of Digital Forensics, recognizing the constant evolution of the technological landscape in recent years. The authors' paper introduced and discussed several solutions and techniques to enhance better collection, analysis, preservation and presentation in the face of the challenges of Digital Forensics.

Yaacoub *et al.* shed light on the current landscape of cyber attacks on IoT and the associated challenges. The authors not only draw conclusions and advocate for further research in the realm of smart automated evidence detection tools, incorporating machine learning techniques but also underscore the necessity for increased investment in the field of forensics. They emphasize that a greater commitment to resources is essential for researchers to effectively address current demands and challenges in the field.

Consistent with the potential of machine learning, Dey *et al.* [8] devised a way to infer the number of people in a specific room of a smart environment using machine learning. In their work, they constructed a model that utilizes temperature, CO_2 levels, air volume, and air conditioning temperature data for their analysis. The authors successfully predicted real-time room occupancy in a building at the University of Washington with up to 95% accuracy. To enhance data analysis, they also explored the use of parallel processing techniques for data analysis and normalization of collected data. The utilization of Data Stream Processing (DSP) techniques proved to be highly beneficial due to the abundance of data ingested by the application, enabling real-time data processing.

3   Study Description

To assess the feasibility of detecting specific patterns in the highly heterogeneous IoT infrastructure within a large dataset, this research aims to assist in the investigation of crimes occurring in physical environments involving the IoT. More specifically, the study compares different machine learning models to estimate the number of occupants in a smart environment, utilizing a variety of heterogeneous sensors.

3.1   Dataset

The primary dataset for this study was originally collected and compiled by Adarsh *et al.* [23]. These data provide a valuable source of information for the analysis conducted. The dataset comprises data from five different types of non-invasive sensors: temperature, luminosity, CO_2, as well as sound and passive infrared (PIR) sensors. This diversity of data offers a comprehensive and detailed insight into the environment analyzed, allowing a more accurate and comprehensive analysis of occupancy patterns.

One of the significant challenges we faced was data availability, as datasets that met the specific criteria required for this research were limited. In addition, ensuring data quality was a meticulous process as we needed to address inconsistencies, errors, and missing values within the dataset. Additionally, the diversity of the data set in terms of sensor types, environmental conditions, and occupancy patterns had to be evaluated, as a more diverse dataset is essential for a comprehensive analysis. Although this dataset proved invaluable for our study, these challenges underscore the importance of having access to more extensive and diverse datasets in the field of occupancy analysis, while also highlighting areas that may benefit from improvements in future research endeavors.

3.2   Exploratory Data Analysis

The analysis of sensor time series graphs, as presented in Fig. 1, revealed that data from the luminosity, temperature, sound, PIR, and CO_2 sensors provide an indication of the number of occupants in the room.

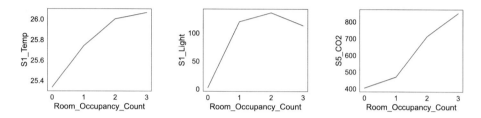

Fig. 1. Representation of sensor data related to the number of occupants.

Furthermore, Fig. 2 allows us to visualize the correlation between variables in the database utilized. We observe that the *S1_Temp* and *S5_CO2* sensors exhibit a strong correlation, indicating that temperature changes may be related to variations in carbon dioxide CO_2 levels. On the other hand, we notice that the *S1_Temp* and *S4_Light* sensors do not exhibit a strong relationship, suggesting that temperature changes are not directly related to variations in luminosity.

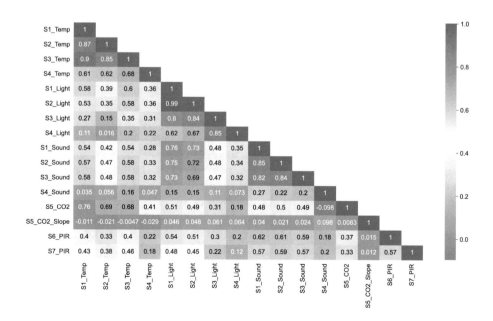

Fig. 2. Correlation analysis between independent and target variables.

A significant correlation between variables can indicate a direct relationship between them, which is useful for prediction and pattern identification. This information is particularly relevant in forensic analyses, where the goal is to uncover specific patterns within the heterogeneity of the IoT.

Identifying highly correlated variables allows us to take advantage of this information to enhance forensic analysis. These variables can provide valuable information on interactions among connected devices and help identify potential evidence during forensic investigations. Understanding the correlation between variables allows one to focus analysis efforts on the most relevant aspects, facilitating the detection of patterns and suspicious behaviors.

By understanding the nature of the data and the correlation among the variables, we can identify which oversampling or undersampling techniques are most suitable for the given dataset [20], thus improving the effectiveness of forensic analyses in IoT environments. However, before applying oversampling or undersampling techniques, it is crucial to detect and address outliers in the datasets.

4   Outlier Detection and Treatment

Outlier detection and treatment are crucial steps in data analysis. Outliers are atypical values that deviate significantly from the pattern of data distribution. These outlier observations may arise due to measurement errors, data flaws, or rare and relevant events. The presence of outliers can distort descriptive statistics and negatively impact the performance of machine learning models [11].

A commonly used method for identifying outliers is the Interquartile Range (IQR), which uses the difference between the third quartile (Q3) and the first quartile (Q1) of a distribution. The IQR represents the range of central data used to define a range within which most of the data is considered normal. Values below Q1 − 1.5 * IQR or above Q3 + 1.5 * IQR are considered outliers [2]. Once identified, there are different approaches to dealing with outliers. In this work, we chose to adopt the approach of replacing outliers in the dataset with the column median. This choice was made because outliers can distort statistical analysis and harm the performance of machine learning models. By replacing these outlier data points, our objective is to ensure a more accurate and reliable representation of patterns within the data, potentially leading to more consistent and reliable outcomes in the analysis and prediction of room occupancy.

Figures 3 illustrate the comparison between certain classes that contain outliers and the same classes after the application of the IQR method (interquartile range). It is evident that after removing outliers using this method, the data distribution became more homogeneous and devoid of outlier values. This indicates that the IQR method was effective in identifying and removing outliers, contributing to a more robust and accurate data analysis [18].

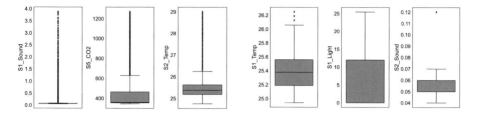

Fig. 3. Outlier analysis before and after the application of IQR.

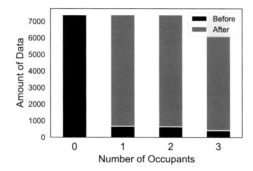

Fig. 4. Dataset before and after SMOTE.

4.1 Data Preprocessing

Data Preprocessing techniques aim to balance the distribution of classes, ensuring that all classes have appropriate representation for training machine learning models. The dataset used consists of more than 10,000 records and 16 columns, each representing data from a specific sensor. However, the dataset was observed to exhibit a significant class imbalance, as illustrated in Fig. 4. Such an imbalance can hinder the performance of certain machine learning models [16]. To address this issue, we chose to employ the oversampling technique known as **SMOTE** (Synthetic Minority Oversampling Technique) [5].

When SMOTE is applied to the dataset, synthetic examples are generated along the segments connecting the nearest neighbors of the minority class. The number of synthetic examples to be generated is determined by a parameter called k, which represents the number of nearest neighbors to consider. The choice of these neighbors is made at random to ensure diversity in the synthetic examples generated [5]. As illustrated in Fig. 4, it can be seen that, after applying SMOTE, the training sample has been balanced so that the minority class matches the majority class in terms of the number of examples.

4.2 Data Normalization

Normalization is a process in which the attributes of a dataset within a model are adjusted or rearranged to enhance the coherence and consistency of the data.

This technique helps in the flexibility of the data, allowing them to be compared and related more efficiently [22]. Through data normalization, redundancy and duplication of information are reduced, avoiding contradictions in the data and rendering it more reliable. This step plays a crucial role, as it classifies and organizes the data in a standardized manner [13].

Data normalization using the **MinMaxScaler** technique is one of the most well-known approaches to data standardization. In this technique, each element of the dataset has its minimum value transformed to 0, the maximum value is transformed to 1, and the remaining values are adjusted to a decimal scale between 0 and 1. This transformation ensures that all elements fall within the same range, facilitating data comparison and analysis [22].

4.3   Model Evaluation

To assess the effectiveness of the models used, performance metrics will be used. In this study, precision, F1 score, accuracy, and recall metrics will be adopted, allowing for a comprehensive evaluation of the model's ability to fit the data and perform the classification task adequately.

To evaluate the models trained with real data, a data splitting approach was used, dividing the data into training (90%) and validation (10%) sets. As illustrated in Fig. 5, this study allocated 10% of the data from the dataset for the final evaluation of the models. These data remained unchanged, without normalization or SMOTE application. The results presented in Sect. 5 elucidate the comparison between the models when applied to these data.

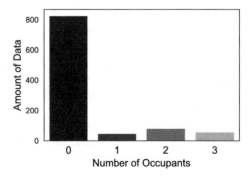

Fig. 5. Real, unaltered sampling for final model validation.

5   Discussion of Results

The results presented in Table 1, shows that models presented satisfactory performance, achieving metrics that exceed 0.90. In particular, the models based on

Random Forest [7] and *XGBoost* [6] exhibited the best results in terms of accuracy, precision, recall, and F1 score. These outcomes indicate that these models are better suited to infer and detect specific patterns in the IoT infrastructure based on the analyzed features. These models demonstrated a superior ability to predict and accurately identify relevant events and behaviors within the context of the IoT, thus contributing to more informed and efficient decision-making.

Regarding the Decision Tree model, a high accuracy rate can be observed across all classes, demonstrating its effectiveness in the classification task. However, there was a higher incidence of errors in predicting the class representing the presence of 2 people in the room, totaling 5 incorrect cases. Figure 6 shows the confusion matrix resulting from applying this model to the analyzed data, providing a visual representation of the performance of the classification algorithm. It shows the distribution of predicted classes in relation to actual classes; for example, in the second class, there were 47 correct matches between predicted and actual labels, with only one error out of a total of 48 labels.

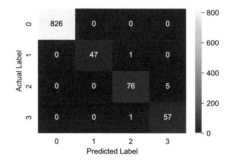

Fig. 6. Confusion Matrix of *Decision Tree.*

Regarding the Random Forest machine learning model, the attributes that exhibited the highest importance were luminosity, sound, and temperature, as can be observed in Fig. 7. These variables had a significant impact on the model's decision-making process, indicating their pivotal role in pattern detection and inference related to the IoT infrastructure. This information is valuable in helping us to understand which specific features were of greater importance.

However, the AdaBoost and Naive Bayes models showed less satisfactory results. This performance difference can be attributed to various factors. In the case of AdaBoost, as mentioned earlier, it can be sensitive to imbalanced data or the presence of outliers. Therefore, if the dataset used possesses these characteristics, it may have adversely affected the performance of the AdaBoost model.

Hyperparameters Tuning. In the realm of machine learning, hyperparameter tuning is the fine-tuning of the settings of a powerful instrument that aims to infer room occupancy from data. These settings, known as hyperparameters,

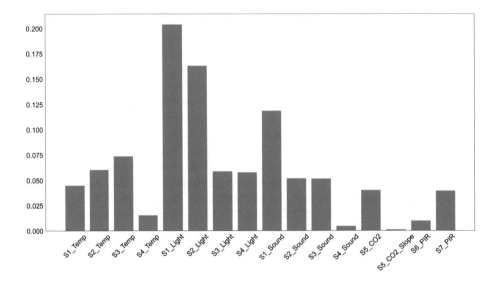

Fig. 7. Importance of Classes for the Random Forest *Random Forest*.

Table 1. Comparison of the Analyzed Models

Model	Accuracy	F1-Score	Precision	Recall
Decision Tree	0.993	0.975	0.976	0.975
Random Forest	0.994	0.978	0.977	0.979
SVM	0.979	0.923	0.924	0.922
KNN	0.973	0.901	0.899	0.904
XGBoost	0.995	0.981	0.981	0.982
Naive Bayes	0.945	0.787	0.799	0.804
AdaBoost	0.859	0.610	0.635	0.632

are the keys to orchestrating the performance of our machine learning models, specifically tailored to the task of inference of room occupancy.

Our primary goal in hyperparameter tuning is to identify the ideal combination of hyperparameters that results in the highest accuracy and reliability in inferring room occupancy. This optimal configuration ensures that our model excels not only in training but, most importantly, in making accurate real-time predictions. However, the computational resources required to navigate a complex landscape of possibilities are substantial. Therefore, we must tread carefully to avoid overfitting our model to the training data.

To address these challenges, we will exclusively employ Grid Search and Random Search to fine-tune our hyperparameters. Grid Search systematically explores a predefined hyperparameter grid, while Random Search randomly samples hyperparameters from specified ranges.

Hyperparameter tuning, particularly through Grid Search and Random Search, is a dynamic process where the right combinations create a powerful model that accurately and reliably infers room occupancy. It is a fundamental part of our journey to harness the full potential of machine learning in the context of room occupancy inference. For the hyperparameter tuning process, we opted for Random Search, primarily because of its superior performance in efficiently exploring the hyperparameter space. Additionally, our focus for hyperparameter tuning was directed at both the Random Forest and XGBoost models.

Table 2. Hypertuned models

Model	Accuracy	F1-Score	Precision	Recall
Random Forest	0.994	0.975	0.977	0.979
XGBoost	0.995	0.978	0.981	0.981

In our research, while rigorously optimizing the hyperparameters for our XGBoost model, we observed a marginal decrease in model performance, compared to the untuned baseline model, as summarized in Table 2. This phenomenon is not uncommon in the field of machine learning and can be attributed to various factors [4]. One possibility is that, in our pursuit of optimizing hyperparameters, the model may have been unintentionally fine-tuned to the validation data to such a degree that it no longer generalized as effectively to unseen data. Additionally, the intricate interactions among hyperparameters can sometimes result in suboptimal combinations that appear to perform well in the validation phase but fail to maintain their effectiveness in real-world data. It is crucial to consider that, due to inherent noise or randomness in real-world datasets, observed fluctuations in model performance can occur. Furthermore, the evaluation metric chosen, f1 score, for optimization, although meticulously selected, may not perfectly align with the real-world goals of our room occupancy inference task. In light of this observation, our study emphasizes the importance of not only the pursuit of optimal model configurations but also the recognition of the broader challenges and trade-offs inherent in machine learning.

6   Limitations

Our research encountered a certain limitation in that our machine-learning models were tested on only our initial dataset. To address this, we took proactive steps to reduce potential biases. We allocated 10% of the initial dataset for testing, allowing us to assess the generalizability and strength of our models beyond the data on which they were trained.

To maximize the accuracy and dependability of our results, we extended our methodology by using cross-validation techniques. This involves dividing the dataset into separate parts, training the models on one part, and then testing them on the other. This iterative process provides a thorough evaluation of the models' performance in different data sets, which helps to more accurately determine their effectiveness in recognizing patterns in various IoT environments. Furthermore, by using cross-validation, our research exceeded the restrictions that came with relying on a single set of tests. It safeguarded against the possibility of overfitting and increased the models' ability to handle data that had not been seen before.

Furthermore, cross-validation is in line with the construct validity framework, which guarantees that our models accurately capture the complex patterns and behaviors common in IoT environments. Finally, by adding cross-validation, we made our findings more reliable, providing a more thorough assessment of how the machine learning models worked with a wider range of data.

7   Conclusion

The results obtained in this study highlight the ability of Random Forest and XGBoost to identify patterns in data collected by IoT devices, solidifying their role in this context. The efficiency of these models in detecting relevant features in the IoT infrastructure demonstrates their potential to assist in criminal investigations within this environment. By applying these models to criminal investigations in IoT environments, it is possible to identify suspicious events and anomalies that may indicate criminal or unauthorized activities. For example, abnormal patterns in brightness, CO_2 levels, or sounds can serve as indications of illicit activities. These models can be adapted and refined over time, enhancing their ability to detect relevant patterns related to crime.

While Random Forest and XGBoost have demonstrated satisfactory efficiency in estimating the number of people in an environment based on data collected from IoT devices, these models must be interpretable. This means that the results should be comprehensible and explainable, allowing investigative professionals and stakeholders to understand how decisions are made and trust the results obtained. In summary, the results underscore that the Random Forest and XGBoost algorithms show the potential to detect room occupancy in IoT environments, highlighting their efficiency for investigation purposes. In this study, both models exhibited the best performance, validating their utility to analyze data from IoT environments.

Although our primary focus lies in examining data produced by IoT devices, the proposed solution demonstrates versatility across different realms of Digital Forensics. This applicability extends beyond the scope of IoT, encompassing various aspects of digital forensics, including areas such as mobile forensics, network forensics, and forensic analysis of multimedia data.

As a future work, we intend to create a new comprehensive dataset. This data collection will be meticulously collected from a single room within the IoT environment, incorporating a wide range of sensory inputs. The primary objective of this data collection effort is to facilitate more thorough testing and validation, offering an improved understanding of how machine learning models can effectively identify patterns and anomalies within the unique context of room dynamics. This specialized dataset is expected to significantly advance forensic IoT analysis, providing further fresh insights into this dynamic field.

Moreover, ongoing research is actively engaged in an extensive investigation into the application of Machine Learning algorithms for pattern detection in the context of criminal investigations within IoT environments. This pattern detection will not be limited solely to room occupancy but will encompass various other machine-learning applications to identify anomalies across the diverse sensor data within these environments.

Acknowledgements. This research was carried out in cooperation with Hewlett Packard Brasil LTDA using incentives of Brazilian Informatics Law (Law n° 8.2.48 of 1991); through subsidies granted by the INCT Call - MCTI/CNPq/CAPES/FAPs No. 16/2014; CAPES (Coordination for the Improvement of Higher Education Personnel - Financing Code 001) and CNPq (National Council for Scientific and Technological Development - 06250/2021-7). Avelino F. Zorzo thanks CNPq/Brazil Grant # 306250/2021-7. Edson OliveiraJr thanks CNPq/Brazil Grant # 311503/2022-5. Roben C. Lunardi thanks IFRS and FAPERGS/Brazil.

References

1. Adedayo, O.M.: Big data and digital forensics. In: 2016 IEEE International Conference on Cybercrime and Computer Forensic (ICCCF), pp. 1–7 (2016)
2. Aggarwal, C.C.: Probabilistic and statistical models for outlier detection. Outlier Analysis, pp. 41–74 (2013)
3. Atlam, H.F., El-Din Hemdan, E., Alenezi, A., Alassafi, M.O., Wills, G.B.: Internet of things forensics: a review. Internet Things **11**, 100220 (2020)
4. Bergstra, J., Bengio, Y.: Random search for hyper-parameter optimization. J. Mach. Learn. Res. **13**(null), 281–305 (2012)
5. Chawla, N.V., Bowyer, K.W., Hall, L.O., Kegelmeyer, W.P.: Smote: synthetic minority over-sampling technique. J. Artif. Intell. Res. **16**, 321–357 (2002)
6. Chen, T., Guestrin, C.: Xgboost: a scalable tree boosting system. In: Proceedings of the 22nd ACM SIGKDD International Conference on Knowledge Discovery and Data Mining, pp. 785–794 (2016)
7. Cutler, A., Cutler, D.R., Stevens, J.R.: Random Forests. Ensemble Machine Learning: Methods and Applications, pp. 157–175 (2012)
8. Dey, A., et al.: Namatad: inferring occupancy from building sensors using machine learning. In: 2016 IEEE 3rd World Forum on Internet of Things (WF-IoT), pp. 478–483 (2016)
9. Garfinkel, S.L.: Digital Forensics innovation: searching a terabyte of data in 10 minutes. DCACM, January 2013
10. Guarino, A.: Digital Forensics as a Big Data Challenge, pp. 197–203. Springer Fachmedien Wiesbaden, January 2013
11. Hair, J.F., Black, W.C., Babin, B.J., Anderson, R.E., Tatham, R.L.: Multivariate data analysis, Upper Saddle River. Multivariate Data Analysis (5th edn.) Upper Saddle River **5**(3), 207–219 (1998)
12. Islam, M.J., Mahin, M., Khatun, A., Debnath, B.C., Kabir, S.: Digital forensic investigation framework for internet of things (iot): a comprehensive approach. In: 2019 1st International Conference on Advances in Science, Engineering and Robotics Technology (ICASERT), pp. 1–6 (2019)
13. Jayalakshmi, T., Santhakumaran, A.: Statistical normalization and back propagation for classification. Int. J. Comput. Theory Eng. **3**(1), 1793–8201 (2011)
14. Kebande, V.R., Karie, N.M., Venter, H.S.: Cloud-centric framework for isolating big data as forensic evidence from iot infrastructures. In: 1st International Conference on Next Generation Computing Applications (NextComp), pp. 54–60 (2017)

15. Kebande, V.R., Ray, I.: A generic digital forensic investigation framework for internet of things (iot). In: IEEE 4th International Conference on Future Internet of Things and Cloud (FiCloud), pp. 356–362 (2016)
16. Kim, J., Kim, J.: The impact of imbalanced training data on machine learning for author name disambiguation. Scientometrics **117**(1), 511–526 (2018)
17. hoon Kim, T., Ramos, C., Mohammed, S.: Smart city and iot. Future Generation Comput. Syst. **76**, 159–162 (2017)
18. Leys, C., Ley, C., Klein, O., Bernard, P., Licata, L.: Detecting outliers: do not use standard deviation around the mean, use absolute deviation around the median. J. Exp. Soc. Psychol. **49**(4), 764–766 (2013)
19. Mohammed, H.J., Clarke, N., Li, F.: An automated approach for digital Forensic analysis of heterogeneous big data. J. Digital Forensics, Secur. Law, January 2016
20. Mohammed, R., Rawashdeh, J., Abdullah, M.: Machine learning with oversampling and undersampling techniques: overview study and experimental results. In: 2020 11th International Conference on Information and Communication Systems (ICICS), pp. 243–248 (2020)
21. OliveiraJr, E., Zorzo, A.F., Neu, C.V.: Towards a conceptual model for promoting digital Forensics experiments. Forensic Sci. Int. Digit. Invest. **35**, 301014 (2020). https://www.sciencedirect.com/science/article/pii/S2666281720301530
22. Raju, V.G., Lakshmi, K.P., Jain, V.M., Kalidindi, A., Padma, V.: Study the influence of normalization/transformation process on the accuracy of supervised classification. In: 2020 Third International Conference on Smart Systems and Inventive Technology (ICSSIT), pp. 729–735. IEEE (2020)
23. Singh, A.P., Jain, V., Chaudhari, S., Kraemer, F.A., Werner, S., Garg, V.: Machine learning-based occupancy estimation using multivariate sensor nodes. In: 2018 IEEE Globecom Workshops (GC Wkshps), pp. 1–6 (2018)
24. Yaacoub, J.P.A., Noura, H.N., Salman, O., Chehab, A.: Advanced digital forensics and anti-digital forensics for iot systems: techniques, limitations and recommendations. Internet Things **19**, 100544 (2022)
25. Yaqoob, I., Hashem, I.A.T., Ahmed, A., Kazmi, S.A., Hong, C.S.: Internet of things forensics: Recent advances, taxonomy, requirements, and open challenges. Futur. Gener. Comput. Syst. **92**, 265–275 (2019)

Advancing Cardiovascular Imaging: Deep Learning-Based Analysis of Blood Flow Displacement Vectors in Ultrasound Video Sequences

Ouissal Kriker[1,2]([envelope]), Asma Ben Abdallah[1], Nidhal Bouchehda[3], and Mohamed Hedi Bedoui[1]

[1] Laboratory of Technologies and Medical Imaging, LR12ES06, Monastir, Tunisia
Krikerouissao@gmail.com
[2] Faculty of Science, University of Monastir, Monastir, Tunisia
[3] Department of Cardiology, Fattouma Bourguiba University Hospital, Monastir, Tunisia

Abstract. Analysis of cardiac hemodynamics from Doppler echocardiography videos allows (i) precise diagnosis of circulatory disorders, (ii) assessment of cardiovascular function and (iii) orientation of medical intervention. The analysis of cardiac hemodynamics is based on determining the movement of blood between successive images. The used techniques fall into two categories: those estimating displacement vectors for each region of interest and those that operate at pixel level. The two main difficulties lie, on the one hand, in the limited availability of cardiac ultrasound data due to confidentiality restrictions and, on the other, in the complexity of the data annotation task, which requires specialist expertise and is time-consuming.

Our proposed approach addresses this limitation by proposing a new algorithm for estimating blood displacement vectors based on supervised deep learning. Our contribution consists in including an automatic annotation step for the video database in order to use the labelled images as ground truth. This approach includes three main stages: (1) data preparation and selection of regions of interest (ROIs), (2) automatic annotation using the optical flow method and (3) motion estimation using deep learning (DL) models. This new algorithm has been validated on a Doppler echocardiography video base. The performance of the automatic annotation achieves AEE = 1.36 and the DL-based displacement estimation optimize the AEE value up to 0.013.

Our novel approach holds considerable potential for accurately estimating and visualizing flow fields in clinical context. With an encouraging and a promising outcome, providing a constructive foundation for further investigation and inspiring the exploration of new perspectives for the diagnosis and treatment of cardiovascular diseases.

Keywords: Cardiac hemodynamics · Motion Tracking · Motion estimation · Flow fields · Flow displacement detection · deep learning (DL)

© The Author(s), under exclusive license to Springer Nature Switzerland AG 2024
Á. Rocha et al. (Eds.): WorldCIST 2024, LNNS 985, pp. 115–125, 2024.
https://doi.org/10.1007/978-3-031-60215-3_12

1 Introduction

Cardiac hemodynamics is the study of blood flow motion through the heart, including the pumping and circulatory mechanisms [1]. This discipline includes phenomena related to cardiac function, such as the contraction of the ventricles, the movement of the heart valves, the regulation of blood pressure and blood circulation. Cardiac hemodynamics issues can precipitate consequential health complications including high blood pressure, heart failure and coronary artery disease. Understanding cardiac hemodynamics is essential for assessing cardiac health, diagnosing cardiac pathology and guiding therapy. A better interpretation of this hemodynamics can be obtained by analyzing blood motion from Doppler echocardiography sequence images.

Typically, blood motion analysis methods estimate blood displacement between two consecutive images of the same flow plane. Both images must be taken at the same height and angle. Motion estimation methods fall into two categories: (1) those estimating displacement vectors for each region and (2) those estimating displacement vectors for pixel-level [2].

The main categories of methods reported in the literature are differential methods, region-based methods, feature-based methods, and CNN-based methods [2].

- Differential methods estimate spatial derivatives of pixel intensities between successive images, assuming constant brightness [3,4].
- Region-Based Methods separate video sequence images into windows and calculate the correlation between each pair to determine the displacement vector [3, 5], as seen in Particle Image Velocimetry (PIV) [6]. It involves the injection of tracer particles into the liquid (blood) and the tracking of its movement to calculate flow velocity and direction (Fig. 1).
- Feature-Based Methods: Features such as gradients, corners and patterns are extracted from successive images over a period of time and then combined to produce a motion representation [7], i.e. the Lucas-Kanade method [8].
- CNN-based methods extract deep features from the input images. These features can be incorporated into standard optimization algorithms to compute the displacement, or used in an end-to-end feature matching method [9] including FlowNet [10] and the deep dilated residual network (RAFT) [11].

Optical flow is a method for computing displacement vectors between successive pairs of images (at time t and t + 1) obtained from a video input, as demonstrated in Fig. 2. This technique encompasses all the aforementioned methods [12]. The displacement field can be estimated sparsely or densely. In the first case, the motion is calculated for a given subset of points (or pixels). These points are usually chosen at strategic locations in the image, for example, using corner or feature detection algorithms. This approach considerably reduces the amount of computation required, but gives a less detailed representation of motion compared to dense optical flow. In the second case, dense motion estimation means that the displacement is calculated for each pixel. This approach provides a detailed representation of pixel motion in the image. The most widely recognized dense estimation algorithms include: Horn-Schunck [4], Gunnar Farneback [13].

Fig. 1. A region-based approach of tracking using a correlation-based method [26].

Fig. 2. Pixel-based approach to track motion: (a), (b) two consecutive images of RubberWhale sequence on Middlebury benchmark [14], (c) vector plot of ground-truth stream [27].

Recently, innovative techniques based on convolutional neural networks (CNNs) have been developed to estimate fluid motion using images. CNNs used for optical flow promise faster results than conventional methods [15]. The multilayer architecture of CNNs facilitates the extraction of deep and abstract features, while their ability to model complex transformations makes them more powerful as motion analysis tools. They can be classified as networks based on U-networks (U-Net) or networks based on spatial pyramids. Using U-Net-based networks, Dosovitskiy et al. [10] were among the pioneers to use convolutional neural networks (CNNs) to learn the optic flow field directly from annotated synthetic data. They developed two networks: FlowNetS and FlowNetC. Both architectures are based on the encoder-decoder model, also known as U-Net. For FlowNetS, the network input is made up of two adjacent images that are superimposed. The encoder has consecutive convolution layers, with some having a reduction factor of 2, which helps deduce the feature map scale as we go deeper into the network. Finally, at the end of the encoder, the scale is reduced to 1/64 of the initial input scale. However, FlowNetC presents a minor alteration. The input is separated into two branches whose outcomes are then entered in a correlation layer which can determine the matching cost of the features. FlowNetS and FlowNetC's decoder part is identical as it consists of a sequence of deconvolution layers. All processes have a factor of two expansions, thus enabling enlargement of the feature maps scale. After the deconvolution process, the network generates the flow field. Lee et al. [16] combined cross-correlation techniques with the power of deep learning (PIV-DCNN). For this, they used a cascade regression architecture to perform PIV estimation. The input of the PIV-DCNN model is a two-channel image with a displacement vector as output. A few works pertaining to fluid dynamics have focused on estimating fluid motion from particle images [16-19]. Ilg et al. [20] stacked U-Net architectures to create a large network called FlowNet2.0. The accuracy of optical flow estimation improves considerably thanks to the stacked iteration of sub-networks. However, model size and execution time increase significantly. Based on a basic encoder-decoder architecture, many excellent implementations have been proposed [19-23]. Meanwhile, using Networks based on spatial pyramid structures, Ranjan et al. [24] developed a spatial pyramid network (SPyNet) based on the coarse-to-fine variational method for managing large displacements. A convolutional network is trained to compute the optical flow at each level of the pyramid, which is subsequently sampled at the next level to guide the deformation from the second image to the first. The model has considerably fewer parameters than FlowNet and yields more accurate results. However, the accuracy of the aforementioned model is inferior to FlowNetS. A modified spatial pyramid network, named PWC-Net, was proposed by Sun et al. [25] using a feature warping operation and a cost volume layer that calculates the matching cost at each pyramid level. Furthermore, Teed et al. [11] developed Recurrent All-Pairs Field Transforms (RAFT) for optical flow estimation that utilized a recurrent update operator to share weights between iterations. These CNN-based models require ground truth data to learn the estimation of displacement vectors and the optimization of the model parameters. They have been trained on several synthetic datasets, such as Flying Chairs [10], MPI Sintel [28], KITTI [29] and Middelebury [14]. Despite the significant progress made by motion detection models, they have certain limitations such as: (1) the difficulty detecting fast movements or movements with occlusions, (2) significant resources and annotated data (i.e. determination of displacement vector for each pixel of the different pairs of successive images extracted from the input video) which is a tedious and time-consuming process.

This problem is much more complex in the case of diagnosing cardiac pathologies using ultrasound images due to the lack of annotated image databases. To overcome these limitations, this paper proposes a novel deep learning (DL)-based optical flow approach for blood motion estimation composed of three steps: (i) pretreatment and region of interest selection (ROI), (ii) Data annotation based on optical flow, and (iii) Deep based method for blood displacement vectors estimation (Fig. 3). The approach was validated to estimate the blood displacement vectors in Doppler echocardiography videos.

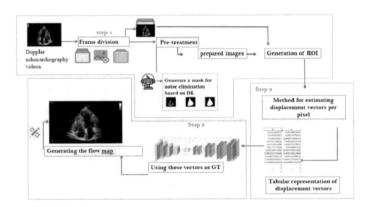

Fig. 3. Pipeline of the DL-based blood motion estimation method.

2 Materials and Methods

2.1 Materials

Our cardiovascular imaging dataset consists of 39 videos acquired during Doppler echocardiographic examinations of 15 patients of whom 3 were pathological cases. These videos have an average duration of 2 s with an image frequency of 60 images/sec. The Doppler echocardiographic videos were provided by the Cardiology Department of Fattouma Bourguiba Teaching Hospital, Monastir.

2.2 Methods

We propose a new optical flow approach based on deep learning (DL) for estimating blood displacement vectors. It consists mainly of three steps: (i) Pretreatment and Region of interest selection (ROI), (ii) Automatic data annotation based on optical flow to tackle the lack of annotated data using dense optical flow method and (iii) deep based estimation of blood displacement.

A) Pretreatment and Region of interest selection (ROI)

Doppler echocardiography videos were used as input, which were initially segmented into frames. The number of frames is dependent on the initial acquisition frequency. The images were resized to 512×512. In total, the dataset contains 4221 images. In order to focus on the analysis of blood in the heart, we generated the region of interest (ROI) including the heart cavities. A DL-based segmentation model was used, more specifically a U-NET network to generate a mask to delimit our ROI, then the final ROI image was obtained using bitwise techniques that extract specific regions of interest from images by using masks [30]. (Fig. 5 (d)). As hyper parameters, we used the ReLU activation function, the Adam optimizer, with a learning rate of 0.001, over 32 epochs, and a batch size of 4 pixels. Cross-validation was performed during the training phase and the dataset was divided into three sections: 20% for validation, 20% for testing and 60% for training. We used the Keras and Tensor flow libraries during implementation.

B) Data automatic annotation based on dense optical flow

To overcome the lack of annotated data, we proposed an automatic annotation method that estimates blood displacement via a dense optical flow named Farneback method [13]. The delimited regions (ROI) are used as input. A pairwise image processing (two successive images) was achieved to calculate the speed and direction of the flow fields. The Farneback method is based on creating a pyramid of the input images, where each level represents the image at a reduced scale. Then displacement vectors were calculated by a polynomial approximation of motion between the two successive images at each level. These calculated vectors were interpolated in order to obtain the overall displacement vectors for all pixels in the original images.

C) Deep based method for blood displacement vectors estimation

Our goal was to estimate displacement vectors using a supervised DL-based approach. The difficulty of annotating our ultrasound database and the lack of expertise we overcome by adopting the dense optical flow technique as an automatic annotation method to prepare the GT. The input of the FlowNetS model consists of image pairs and GT flows. In order to estimate the blood displacement vectors (Fig. 4), several hyper parameters were thoughtfully chosen to enhance the model's performance. We initially opted for incorporating the LeakyReLU activation function to introduce non-linearity into our neural network. For optimization, we chose the Adam optimizer with an initial learning rate of 0.001. Our training schedule runs for 50 epochs. Furthermore, to ensure the reliability of our model's performance evaluation, we conducted a 5-fold cross-validation. To adjust the learning rate during training, we employed a 10-epoch step size and a 0.1 reduction factor with the StepLR learning rate scheduler.

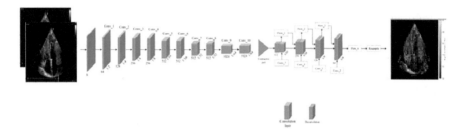

Fig. 4. FlowNetS architecture for blood displacements estimation using Doppler echocardiography dataset.

2.3 Metrics

A) Region of interest selection

In order to evaluate the selection of ROI accuracy, the following metrics were used: Dice coefficient, Precision and Recall. Dice Coefficient is a widely used similarity index for segmentation evaluation, while Precision measures the percentage of true positive values that are correctly identified. Accuracy measures the percentage of positive predictive value and F1 scores the overlap between predicted segmentation and ground truth (Table 1).

B) Data automatic annotation based on dense optical flow

Optical flow estimation is a challenging task due to the presence of noise, occlusions, and motion blur in real-world images. To evaluate the accuracy of our automatic annotation method we used as metrics:

– Average Endpoint Error (AEE) is defined as the Euclidean distance between the estimated optical flow vector and the ground truth optical flow vector for each pixel.

$$APE = \frac{1}{N}\sum_{i=1}^{N}\sqrt{[(U_{EST} - U_{GT})^2 + (V_{EST} - V_{GT})^2]} \tag{1}$$

where N is the number of image pixels, U_{EST} and V_{EST} are the estimated optical flow components for pixel i, and U_{GT} and V_{GT} are the ground truth optical flow components for pixel i.

– Root Mean Square Error (RMSE is defined as the square root of the mean of the squared errors between the estimated optical flow vector and the ground truth optical

Table 1. Segmentation metrics equation.

Dice coefficient	Precision	Recall	F1 score	Accuracy
$\frac{2*TP}{2*(TP+FP+FN)}$	$\frac{TP}{TP+FP}$	$\frac{TP}{TP+FN}$	$2 \times \frac{précision \times rappel}{précision + rappel}$	$\frac{TP+TN}{nbredespixels}$

flow vector for each pixel.

$$RMSE = \sqrt{\frac{1}{N}\sum_{i=1}^{N}[(U_{EST} - U_{GT})^2 + (V_{EST} - V_{GT})^2]} \quad (2)$$

where N is the number of image pixels, U_{EST} and V_{EST} are the estimated optical flow components for pixel i, and U_{GT} and V_{GT} are the ground truth optical flow components for pixel i.

3 Experiment and Results

3.1 Region of Interest Selection Results

As input, an image and its GT were provided to the U-net network in order to generate the ROI mask. The result is depicted in Table 2. An example of ROI generation is shown in Fig. 5 where (a) the original image, (b) the GT image mask (c) is the predicted mask image by U-net and (d) the result of the generation of our ROI.

Table 2. Region of interest selection metrics result.

Methods	Dice_coef	F1	Accuracy	Precision	Recall
U-net	0.9885	0.9887	0.9895	0.9928	0.9823

The provided U-net results indicated that the U-net model achieved good performance on the segmentation task. The dice coefficient, specificity, F1 score, accuracy, precision, and recall were all high, indicating that the model was able to segment the ROI accurately.

(a)   (b)   (c)   (d)

Fig. 5. Region of interest selection result.

3.2 Results of the Dense Optical Flow Automatic Annotation

Using the dense optical flow technique for automated annotation of blood displacements, a displacement field was estimated for each pixel as shown in (Fig). These displacements were used as the GT for the DL estimation of displacement vectors in the FlowNetS architecture. To evaluate the performance of the automatic annotation, we performed a comparative study between predicted flows with the provided expert GT. At this stage, it is important to note that the expert GT was originally estimated using a PIV-based method and involving subdivision into 4 × 4 windows.

For this, the displacement vectors matrix dimension was reduced by subdividing images into 4 × 4 size windows. The average of the 16 displacement vectors pixels in each window determines the value of its center, which replaces the entire window. The results are shown in Table 3 show that the displacement values estimated by the dense optical flow are very close to the GT data with AEE = 1,36 (Fig. 6(a)(b)). These results were very encouraging compared with the values ranging from 1.4 to 24 reported in the literature [2].

3.3   Results of the DL Blood Displacement Vectors Estimation

The FlowNetS model was trained to predict blood displacement vectors using pairs of successive images from the Doppler ultrasound database and their GT. Figure 6(d) illustrates the estimated displacement. To evaluate the performance of these predicted displacements vectors, a comparative study between DL predicted with the provided per-pixel optical flow GT. Our results revealed the efficiency of the adopted approach to detect blood motion estimation, owing to its ability to capture hidden patterns and optimize the AEE value up to 0,013. The achieved outcome was both promising and very encouraging, providing a constructive basis for further research and encouraging to explore other approaches.

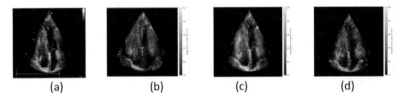

(a)	(b)	(c)	(d)

Fig. 6. Illustration of Blood displacement vectors estimated by different methods on two pairs of images at time t and t + dt from the Private database. Where (a) the GT estimated displacement by our expert, (b) the GT estimated displacement by optical flow technique followed by the replacement of 4 × 4 size windows with their average size (c) the GT per-pixel optical flow technique displacement estimation, (d) the predicted displacement vectors with FlowNetS.

Table 3. Results of the displacement vector estimation.

	RMSE	AEE
Optical flow	2.8216	1.3615
FlowNetS	0.0756	0.013

4   Conclusion

Blood displacement methods stand as a promising technology with a wide potential for medical applications. CNN-based methods are currently the most popular and efficient methods for determining displacement. They have several advantages, including their flexibility. In this work, we leverage the benefits of the CNN models by proposing a new algorithm for estimating blood displacement vectors based on supervised deep learning. Our contribution consists in including an automatic annotation step for the video database in order to use the labelled images as ground truth. This approach comprises three main stages: data collection, pretreatment and selection of regions of interest (ROIs) based DL, (2) automatic annotation using the per-pixel optical flow method and (3) displacement estimation using deep learning (DL) model, precisely FlowNetS. Our approach was validated on a Doppler echocardiography videos dataset. The automatic annotation achieved an AEE of 1.36, while the DL-based displacement estimation achieved an AEE of 0.013. These results were promising and encouraging, as they demonstrated the effectiveness of our approach in detecting complex transformations, making it a powerful tool.

Future research will focus on developing methods that combine the advantages of CNN-based and traditional methods. These methods may be more accurate and robust for challenging scenarios, such as rapid motion or obstructions.

References

1. Nishimura, R.A., Carabello, B.A.: Hemodynamics in the cardiac catheterization laboratory of the 21st century. Circulation **125**(17), 2138–2150 (2012)
2. Zhai, M., et al.: Optical flow and scene flow estimation: a survey. Pattern Recogn.Recogn. **114**, 107861 (2021)
3. Burton, A., Radford, J.: Thinking in perspective: critical essays in the study of thought processes. Taylor & Francis (2022)
4. Horn, B.K.P., Schunck, B.G.: Determining optical flow. Artif. Intell.. Intell. **17**(1–3), 185–203 (1981). https://doi.org/10.1016/0004-3702(81)90024-2
5. Warren, D.H., Strelow, E.R. (eds.): Electronic Spatial Sensing for the Blind: Contributions from Perception, Rehabilitation, and Computer Vision, vol. 99. Springer Science & Business Media (2013)
6. Atkins, M.D.: Velocity field measurement using particle image velocimetry (PIV). In: Application of Thermo-Fluidic Measurement Techniques: An Introduction, pp. 125–166 (2016). doi: https://doi.org/10.1016/B978-0-12-809731-1.00005-8
7. Wills, J., Belongie, S.: A feature-based approach for determining dense long-range correspondences. In: Computer Vision-ECCV 2004: 8th European Conference on Computer Vision, Prague, Czech Republic, May 11–14, 2004. Proceedings, Part III, 8. Springer, Heidelberg (2004)
8. Ayvaci, A., Raptis, M., Soatto, S.: Sparse occlusion detection with optical flow. Int. J. Comput. VisionComput. Vision **97**, 322–338 (2012)
9. Lai, S.-H., et al. (eds.): Computer Vision–ACCV 2016: 13th Asian Conference on Computer Vision, Taipei, Taiwan, November 20–24, 2016, Revised Selected Papers, Part III. Vol. 10113. Springer (2017)
10. Fischer, P., et al.: Flownet: Learning optical flow with convolutional networks. arXiv preprint arXiv:1504.06852 (2015)

11. Teed, Z., Deng, J.: RAFT: recurrent all-pairs field transforms for optical flow. In: Computer Vision–ECCV 2020: 16th European Conference, Glasgow, UK, August 23–28, 2020, Proceedings, Part II, 16. Springer (2020)

12. Ogle, K.N.: The perception of the visual world. In: Gibson, J.J., Carmichael, L. (ed.) Houghton Mifflin, Boston (1950); 235 pp. $4.00. Science, 113(2940), 535–535 (1951)

13. Farnebäck, G.: Polynomial expansion for orientation and motion estimation. Linköping University Electronic Press (2002)

14. Stanley, N., et al.: Development of 3-D printed optically clear rigid anatomical vessels for particle image velocimetry analysis in cardiovascular flow. In: ASME International Mechanical Engineering Congress and Exposition, vol. 59445. American Society of Mechanical Engineers (2019)

15. Tu, Z., et al.: A survey of variational and CNN-based optical flow techniques. Signal Process. Image Commun. **72**, 9–24 (2019)

16. Lee, Y., Yang, H., Yin, Z.: PIV-DCNN: cascaded deep convolutional neural networks for particle image velocimetry. Exp. Fluids **58**, 1–10 (2017)

17. Rabault, J., Kolaas, J., Jensen, A.: Performing particle image velocimetry using artificial neural networks: a proof-of-concept. Meas. Sci. Technol. **28**(12), 125301 (2017). https://doi.org/10.1088/1361-6501/AA8B87

18. Cai, S., et al.: Deep-PIV: a new framework of PIV using deep learning techniques. In: Proceedings of the 13th International Symposium on Particle Image Velocimetry—ISPIV, 201

19. Zhang, M., Piggott, M.D.: Unsupervised learning of particle image velocimetry. In: Lecture Notes in Computer Science (including subseries Lecture Notes in Artificial Intelligence and Lecture Notes in Bioinformatics), 12321 LNCS, 102–115 (2020). https://doi.org/10.1007/978-3-030-59851-8_7

20. Ilg, E., et al.: FlowNet 2.0: evolution of optical flow estimation with deep networks. In: Proceedings of the IEEE Conference on Computer Vision and Pattern Recognition (2017)

21. Zhao, S., Li, X., El Farouk Bourahla, O.: Deep optical flow estimation via multi-scale correspondence structure learning. In: IJCAI International Joint Conference on Artificial Intelligence, pp. 3490–3496 (2017). https://doi.org/10.24963/ijcai.2017/488

22. Vaquero, V., Ros, G., Moreno-Noguer, F., Lopez, A.M., Sanfeliu, A.: Joint coarse-and-fine reasoning for deep optical flow. In: Proceedings - International Conference on Image Processing, ICIP, 2017-September, pp. 2558–2562 (2018). https://doi.org/10.1109/ICIP.2017.8296744

23. Xiang, X., Zhai, M., Zhang, R., Qiao, Y., El Saddik, A.: Deep optical flow supervised learning with prior assumptions. IEEE Access **6**, 43222–43232 (2018). https://doi.org/10.1109/ACCESS.2018.2863233

24. Ranjan, A., Black, M.J.: Optical flow estimation using a spatial pyramid network, pp. 4161–4170 (2017). https://github.com/anuragranj/spynet. Accessed 03 Nov 2023

25. Sun, D., et al.: Pwc-net: Cnns for optical flow using pyramid, warping, and cost volume. In: Proceedings of the IEEE Conference on Computer Vision and Pattern Recognition (2018)

26. Anand, S., Poovitha, R., Nikhila, K.: Enhancement of particle image velocimetry images. arXiv preprint arXiv:2004.10498 (2020)

27. Yedjour, H.: Optical flow based on Lucas-Kanade method for motion estimation. In: Artificial Intelligence and Renewables Towards an Energy Transition 4. Springer International Publishing (2021)

28. Butler, D.J., et al.: A naturalistic open source movie for optical flow evaluation. In: Computer Vision–ECCV 2012: 12th European Conference on Computer Vision, Florence, Italy, October 7–13, 2012, Proceedings, Part VI 12. Springer, Heidelberg (2012)

29. Garcia-Garcia, A., et al.: A review on deep learning techniques applied to semantic segmentation. arXiv preprint arXiv:1704.06857 (2017)
30. Pál, A.: FITSH–a software package for image processing. Mon. Not. R. Astron. Soc. **421**(3), 1825–1837 (2012)

Ethics, Computers and Security

Artificial Intelligence Cyberattacks in Red Teaming: A Scoping Review

Mays Al-Azzawi®, Dung Doan®, Tuomo Sipola$^{(\boxtimes)}$®, Jari Hautamäki®,
and Tero Kokkonen®

Institute of Information Technology, Jamk University of Applied Sciences, Jyväskylä,
Finland
{ab0168,aa7785}@student.jamk.fi,
{tuomo.sipola,jari.hautamaki,tero.kokkonen}@jamk.fi

Abstract. Advances in artificial intelligence are creating possibilities to
use these methods in red team activities, such as cyberattacks. These AI
attacks can automate the process of penetrating a target or collecting
sensitive data while accelerating the pace of carrying out the attacks.
This survey explores how AI is employed in cybersecurity attacks and
what kind of targets are typical. We used scoping review methodology to
sift through articles to find out AI methods, targets, and models that red
teams can use to emulate cybercrime. Out of the 470 records screened, 11
were included in the review. Multiple cyberattack methods can be found
to exploit sensitive data, systems, social media user profiles, passwords,
and URLs. The use of AI in cybercrime to build versatile attack models
poses a growing threat. Additionally, cybersecurity can use AI-based
techniques to offer better protection tools to deal with those problems.

Keywords: artificial intelligence · red team · red teaming ·
cyberattack · cybersecurity

1 Introduction

The landscape of cybersecurity has undergone an enormous change in the last few
years. One phenomenon that stands out is the possibility of artificial intelligence
simulating human behavior. The behavior of artificial intelligence in cybersecurity can lead to dangerous situations in terms of security. Using AI as a method
for attacks has developed in tandem with the development of attack methodologies and AI capabilities. Only a few cases are reported, and simulating human
acts has become more feasible in the last few years.

The term red teaming originates from the military domain as a way to role-
play adversaries or assess vulnerabilities [12]. The term red team also originates
from widely used military symbols such as APP-6 by NATO or MIL-STD-2525 by
U.S. Department of Defense, where the hostile (and suspect) identity is indicated
with a red color [1,15]. In the context of cybersecurity, U.S. National Institute
of Standards and Technology (NIST) defines a red team as follows: *"A group of*

© The Author(s), under exclusive license to Springer Nature Switzerland AG 2024
A. Rocha et al. (Eds.): WorldCIST 2024, LNNS 985, pp. 129–138, 2024.
https://doi.org/10.1007/978-3-031-60215-3_13

people authorized and organized to emulate a potential adversary's attack." [5] The red teams improve enterprise security by demonstrating the impacts of successful attacks [5]. In the context of cybersecurity, the term red team is used in cybersecurity exercises and in security testing. In cybersecurity exercises, red teams (RT) simulate the threat actors of the exercise scenario by executing cyberattacks against blue teams (BT), which are defending their assets [3,9,11,19,24]. In security testing, the red team is the group of security testers.

AI red teaming can be understood as an activity from two different perspectives. Several large technology companies use red teaming to expose weaknesses and vulnerabilities in their systems [18,27]. Another aspect is the use of AI to carry out attacks, which can be targeted against technical systems. On the other hand, in social engineering-type attacks, AI is used as a stepping stone to advanced persistent threat (APT) attacks by searching for suitable victims that can be targeted by AI-generated ghost messages [6,17]. The advantage of AI specifically in such attacks is the ability to enable mass attacks using phishing techniques to open attack vectors to multiple targets instead of manual attacks. For example, AI-generated phishing messages in target language create persuasive attack vectors. AI-based solutions are built to make operations more effective. Automating the process of planning attacks for automated cybersecurity testing scenarios could save time and effort [26]. As new artificial intelligence technologies have become more prevalent, automation is easier to implement, although its impact on work and society should be studied [22].

In order to investigate the use of AI for cyberattacks for red teaming, we carried out a scoping review. To examine how AI can be used for cyberattacks, red team actions, and hacking, our research questions were the following:

- *RQ1:* What AI attack methods are there?
- *RQ2:* What are the targets of such attacks?

Next, this paper describes the used scoping review methodology in Sect. 2, including a figure of the review protocol. The results of the review are presented in Sect. 3 with two tables summarizing the main findings. Finally, a conclusion is provided in Sect. 4.

2   Methodology

We used the scoping review method [14] to search the academic Finna[1] library database and Google Scholar[2] in order to define the scope of our topic. The review considered the following keywords: 'defensive mission', 'AI-enabled cyber operations', 'AI-augmented cyber defenses', 'national defense postures', 'poisoning attacks', 'offensive cyber operations', 'Cyber activities', 'AI cyber operations', 'AI cyber defense', 'AI cyber attack', 'AI red teaming', 'AI-enabled cyber campaigns', and 'cyber attacks'. In the initial stage, we identified 471 articles

---

[1] https://janet.finna.fi/.
[2] https://scholar.google.com/.

(and some book chapters) by screening their titles and abstracts within the 2015–2023 timeframe, found at the time of the research in mid-2023. We included articles written in English with available abstracts. During the second phase of the research, a more involved analysis of these articles was conducted. This analysis included reading the articles closely and concentrating on the topic at hand to precisely determine their content and classify them as directly relevant to addressing the research questions (RQ1, RQ2). We used the following criteria to find answers:

1. Is there a description of an attack method?
2. What was the target of the attacks?
3. How was the attack conducted?
4. What was the cyber-attack methodology used?

The result of the second stage of the research yielded 11 articles related to the subject matter. In the third stage of the research, we composed summaries, which also involved addressing the aforementioned questions when applicable. This comprehensive analysis of the included studies enabled us to gather information on the utilization of AI in red teaming. The review process is detailed by the PRISMA flow chart [13] in Fig. 1.

3   Results

3.1   Attack Methods

The literature review encompassed studies published from 2015 to 2023 in which we identified various cyberattack methods. The following techniques were documented in those studies. *Classification methods:* decision tree, convolutional neural network (CNN), recurrent neural network (RNN), long short-term memory (LSTM), support vector machine (SVM), support vector classification (SVC), deep neural network (DNN), least squares support vector machine (LS-SVM), natural language processing (NLP), one-versus-all (OVA), double deep Q-network (DDQN), advantage actor-critic (A3C) regularized least-squares classification (RLSC), domain generation algorithms (DGA). *Regression methods:* generative adversarial network (GAN), random forest (RF), multilayer perceptron (MLP), gradient boosting regression trees (GBRT), artificial neural network (ANN), logistic regression, generalized likelihood ratio test (GLRT), *Clustering strategies:* k-means clustering, restricted Boltzmann machine (RBM), particle swarm optimization (PSO), genetic algorithm (GA), deep autoencoder (DAE), Lagrangian firefly algorithm (LFA), *Other specific methods:* nonsymmetric deep autoencoder (NDAE), cycle-GAN, combining TensorFlow object detection and a speech segmentation method with convolutional neural network (TOD+CNN), k-nearest neighbors (KNN), reinforcement learning (RL), gray wolf optimization (GWO), random weight network (RWN), ML-based approach named MLAPT, software-defined networking (SDN), and singular value decomposition (SVD).

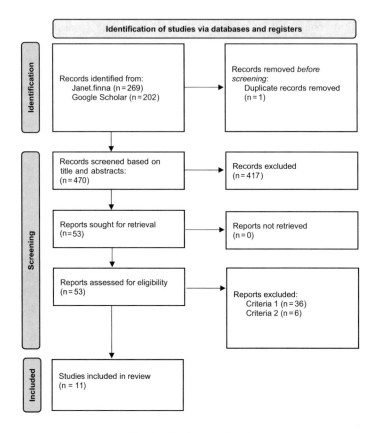

Fig. 1. Review protocol

Among these methods, LSTM was the most frequently used, appearing in 5 of the reviewed articles, while GANs and SVM were employed in 4 studies each. Additionally, CNN, RNN, KNN, MLP, and DNN were each featured in three of the reviewed articles. Other methods were referenced only once or twice. For a list of the attack methods in the reviewed articles, refer to Table 1.

3.2 Attack Targets

Furthermore, we identified common targets that cyberattackers typically aim at (see Table 2 for tabulation of targets), including:

– General data, such as health data, personal data, and sensitive data, including financial and government data, were the most frequently targeted, appearing in 4 of the reviewed articles [2, 20, 25, 28].
– URLs: Attackers also frequently targeted URLs, with 3 instances in the reviewed articles [7, 10, 28].
– Social media user profiles: This category was the target in 2 of the sources [7, 10].

- Passwords: Passwords were a target in 2 sources [7,28].
- Details of systems: Details of systems were targeted in one article [25].

3.3 Summaries

The use of AI has been identified as a cyberattack method and recognized as a potential risk. However, Clinton only presents AI as a hacking method, and we did not find any other specific attack methods [4].

Ward et al. have defined artificial intelligence as a new technology used by hackers and have mentioned "poison" attacks utilizing machine learning algorithms. They also discuss automated vehicles and the potential for high-risk attacks on vehicle systems. However, during their discussion of AI hacking methods, no specific attack methods were mentioned [23].

From 2015 to 2018, the articles about AI-hacking did not mention any attack methods, and targets were mainly data and sensitive data.

Table 1. Methodologies found in the reviewed articles.

Author	Pistono and Yampolskiy	Brundage et al	Kaloudi and Li	King et al	Truong et al	Zouave et al	Yamin et al	Wang et al	Guembe et al	Σ
Year	2016	2018	2020	2020	2020	2020	2021	2022	2022	
Reference	[16]	[2]	[8]	[10]	[20]	[28]	[25]	[21]	[7]	
Classification										
Dec. tree						x				1
CNN					x	x		x		3
RNN			x			x		x		3
LSTM			x		x	x	x	x		5
SVM					x	x	x	x		4
SVC						x		x		2
DNN			x				x	x		3
LS-SVM					x					1
NLP						x				1
OVA						x				1
DDQN							x			1
A3C							x			1
RLSC						x				1
DGA						x				1
Regression										
GANs						x	x	x	x	4
RF						x		x		2
MLP					x	x		x		3
GBRT						x		x		2
ANN					x					1
Log. reg						x				1
GLRT					x					1
Clustering										
k-means			x							1
RBM					x					1
PSO					x					1
GA					x					1
DAE					x					1
LFA					x					1
Other										
NDAE					x					1
CYCLE-GAN									x	1
TOD+CNN									x	1
KNN					x	x			x	3
RL			x							1
GWO					x					1
RWN					x					1
MLAPT					x					1
SDN								x		1
SVD								x		1

Table 2. Attack targets found in the reviewed articles.

Author	Pistono and Yampolskiy	Brundage et al	Kaloudi and Li	King et al	Truong et al	Zouave et al	Yamin et al	Wang et al	Guembe et al	∑
Year	2016	2018	2020	2020	2020	2020	2021	2022	2022	
Reference	[16]	[2]	[8]	[10]	[20]	[28]	[25]	[21]	[7]	
Data, sensitive data		x			x	x	x			4
URLs				x		x			x	3
Social media user profiles				x					x	2
Password						x			x	2
Systems								x		1

Yamin et al. focused on raising awareness about the use of artificial intelligence as an attack method and assessed its impact on military operations. They employed GANs and Nash equilibrium to describe the attack methods. The targets of these attacks included traffic signs, medical image data, facial image data, digital recommendation systems, CT-scan data, speech and audio data, as well as network intrusion detection systems. The attacks were carried out using malicious AI algorithms designed to manipulate data to evade benign AI algorithm classifiers. The methodologies employed in these cyberattacks included DeepHack, DeepLocker, GyoiThon, EagleEye, Malware-GAN, UriDeep, Deep Exploit, and DeepGenerator [25].

The article by Kaloudi et al. investigates AI's threat to SCPS. It explores how AI can be used as a malicious tool, emphasizing its potential to increase attack speed and success rates. Attack methods discussed include k-means clustering, RNN, LSTM, RL, and DNN. Case studies involve k-means clustering for phishing messages, RNN for deceptive reviews, LSTM for phishing URLs, RL for autonomous learning attacks, and DNN for cyberattacks. The paper also examines cyberattack methodologies, including DeepLocker, repurpose attacks, DeepHack, Deep-Phish, review attacks, and SNAP_R [8].

Guembe et al. address the growing concern of AI-powered cyberattacks and provide insights into how AI can be maliciously utilized in such attacks. They employ various attack methods, including CNN, GAN, RNN, LSTM, SVC, SVM, cycle-GAN, TOD+CNN, RF, MLP, GBRT, KNN, and DNN. The targets of these attacks encompass public social media profiles, passwords, and URLs. The attacks are executed through techniques such as password guessing/cracking (brute-force attacks), intelligent captcha manipulation, smart abnormal behavioral generation, AI model manipulation, and the generation of sophisticated fake reviews. The cyberattack methodologies employed by the authors include DeepLocker, DeepHack, PassGAN, and HashCat [7].

Truong et al. provide an insightful overview of how artificial intelligence can be leveraged in cybersecurity, both for offensive and defensive purposes. They employ a diverse set of attack methods, including SVM, RBM, MLP, KNN, CNN, PSO, GA, DAE, ANN, LS-SVM, NDAE, GWO, RWN, LFA, MLAPT, LSTM, and GLRT. The targets of these attacks encompass user identities, financial credentials, and sensitive data from large corporations, security agencies, and government organizations. These attacks serve various purposes, including detecting or categorizing malware, identifying network intrusions, countering phishing and spam attacks, mitigating Advanced Persistent Threats (APTs), and identifying domains generated by domain generation algorithms (DGAs) [20].

Articles in 2020 showed different attack methods, such as GANs, CNN, RNN, LSTM, SVM, and SVC, aimed at attacking sensitive data, social media user profiles, passwords, and URLs.

The article by Zouave et al. explores the possibilities and applications of AI throughout various stages of a cyberattack. The authors employ a wide range of attack methods, including RNN, LSTM, NLP, GAN, KNN, logistic regression, SVC, decision tree, RF, gradient boosting regression tree, SVM, MLP, RLSC, OvA, CNN, and DGA. These attacks target URLs, individuals' personal data in search of relationships, passwords, captchas, and domains. The attacks are executed by creating deceptive URLs to evade automated detection, generating conversations that include harmful links and attachments, attempting password guessing and brute forcing, stealing passwords, solving captchas, and generating numerous random fake domains. The authors utilize cyberattack methodologies such as the DeepPhish algorithm, PassGAN, Torch RNN, Deeptcha, AGDs, and DeepDGA [28].

In the article by Wang et al. the exploration focuses on poisoning attacks in machine learning, particularly within the context of automated vehicles. The authors utilize various attack methodologies harnessing AI techniques. These include deep learning and deep neural networks (DNN), known for their outstanding performance in recognition tasks like image classification and computer vision. Additionally, other methods are discussed, such as Generative Adversarial Networks (GAN), LSTM, SDN, DDQN (Deep Double Q-Network), Advantage Actor-Critic (A3C), SVM (Support Vector Machine), and singular value decomposition (SVD) [21].

The article by Brundage et al. provides a summary of workshop findings and the authors' conclusions on forecasting, preventing, and mitigating the detrimental impacts of malicious AI use. The targets included sensitive information or financial assets of individuals, specific members of crowds, and historical patterns of code vulnerabilities. The attacks were executed through various methods, such as spear phishing attacks, imitation of human-like behavior, facial recognition, the generation of custom malicious websites/emails/links, visual impersonation of another person in video chats, and the use of drones or autonomous vehicles to deliver explosives and cause accidents. Furthermore, the attackers were engaged in discovering new vulnerabilities and developing code to exploit them. However, no specific methods for these activities were mentioned in the report [2].

In their article, King et al. introduced the term "AI-Crime" (AIC) to address two key questions regarding the threats posed by AI in criminal activities and potential solutions to mitigate these threats. However, the article does not specify the methods employed in these AIC activities. The primary target of these activities is social media users, particularly through the use of phishing links [10].

The research paper by Pistono and Yampolskiy focuses on publishing papers related to malicious exploits and discusses the use of software with malicious capabilities, including truly artificially intelligent systems such as artificially intelligent viruses. The paper also introduces the term "Hazardous Intelligent Software" (HIS) to describe the use of intelligence in a malicious context. It highlights that intelligent systems can potentially become malevolent in various ways. However, the paper does not mention specific AI attack methods [16].

4   Conclusion

In today's rapidly evolving digital landscape, cybercriminals are continuously adapting and enhancing their attack strategies, with a particular focus on leveraging AI-driven techniques. Our results indicate that primary targets (RQ2) include personal data as well as sensitive information held by governments, organizations, and individuals, spanning URLs, passwords, and critical systems. Furthermore, the results show that to achieve their malicious goals (RQ1), cybercriminals can exploit a wide array of machine learning methods, falling into distinct categories: Classification, Regression, and Clustering.

These categories contain various technologies used for attacks. *Classification Techniques:* They include a multitude of machine learning algorithms such as decision trees, CNN, RNN, LSTM, SVM, SVC, DNN, LS-SVM, NLP, OVA, DDQN, A3C, RLSC, and DGA. These methods enable cybercriminals to classify and categorize data, often to identify vulnerabilities or potential targets. *Regression Methods:* In this category, we find techniques such as GANs, RF, MLP, GBRT, MLP, ANN, Logistic Regression, and GLRT. These approaches are employed to predict and estimate various variables, ranging from password guessing to system security breaches. *Clustering Strategies:* Cybercriminals also rely on clustering methods such as k-means clustering, RBM, PSO, GA, DAE, and LFA. Clustering helps them identify patterns within data, which can be exploited for nefarious purposes.

Cybercriminals employ sophisticated methodologies like the DeepPhish algorithm, PassGAN, Torch RNN, and Deeptcha. These tools aid them in tasks such as cracking passwords, phishing attacks, and infiltrating secure systems. As the threat landscape continues to evolve, it is imperative for the security research community, government agencies, and cybersecurity experts to remain vigilant and well-prepared against AI-based attacks. Red teaming using these AI-based attacks could reveal vulnerabilities to novel attacks. Effective countermeasures and proactive strategies must be developed to address the growing challenges posed by AI-driven cyberattacks.

Acknowledgements. This research was partially funded by the Resilience of Modern Value Chains in a Sustainable Energy System project, co-funded by the European Union and the Regional Council of Central Finland (grant number J10052). The authors would like to thank Ms. Tuula Kotikoski for proofreading the manuscript.

References

1. Department of Defence Interface Standard, Common Warfighting Symbology. Standard MIL-STD-2525C, United States of America, Department of Defence (2008)
2. Brundage, M., et al.: The malicious use of artificial intelligence: forecasting, prevention, and mitigation. arXiv preprint arXiv:1802.07228 (2018)
3. Brynielsson, J., Franke, U., Tariq, M.A., Varga, S.: Using cyber defense exercises to obtain additional data for attacker profiling. In: 2016 IEEE Conference on Intelligence and Security Informatics (ISI), pp. 37–42 (2016). https://doi.org/10.1109/ISI.2016.7745440
4. Clinton, L. (ed.): Cybersecurity for Business. Kogan Page, London (2022)
5. Computer Security Resource Center (CSRC) of National Institute of Standards and Technology (NIST). The Glossary of Terms and Definitions Extracted Verbatim from NIST's Cybersecurity- and Privacy-Related Publications. https://csrc.nist.gov/glossary/term/red_team. Accessed 15 Sept 2023
6. Ghafir, I., Prenosil, V.: Advanced persistent threat and spear phishing emails. In: Hrubý, M. (ed.) Proceedings of the International Conference Distance Learning, Simulation and Communication 'DLSC 2015', pp. 34–41. University of Defence, Brno (2015)
7. Guembe, B., Azeta, A., Misra, S., Osamor, V.C., Fernandez-Sanz, L., Pospelova, V.: The emerging threat of AI-driven cyber attacks: a review. Appl. Artif. Intell. **36**(1), 2037254 (2022)
8. Kaloudi, N., Li, J.: The AI-based cyber threat landscape: a survey. ACM Comput. Surv. **53**(1), 1–34 (2020)
9. Kick, J.: Cyber exercise playbook (2014). https://www.mitre.org/news-insights/publication/cyber-exercise-playbook. Accessed 15 Sept 2023
10. King, T.C., Aggarwal, N., Taddeo, M., Floridi, L.: Artificial intelligence crime: an interdisciplinary analysis of foreseeable threats and solutions. Sci. Eng. Ethics **26**, 89–120 (2020)
11. Kokkonen, T., Puuska, S.: Blue team communication and reporting for enhancing situational awareness from white team perspective in cyber security exercises. In: Galinina, O., Andreev, S., Balandin, S., Koucheryavy, Y. (eds.) Internet of Things, Smart Spaces, and Next Generation Networks and Systems. LNCS, vol. 11118, pp. 277–288. Springer, Cham (2018). https://doi.org/10.1007/978-3-030-01168-0_26
12. Longbine, D.F.: Red Teaming: Past and Present. School of Advanced Military Studies, Fort Leavenworth (2008)
13. McGowan, J., et al.: Reporting scoping reviews—PRISMA ScR extension. J. Clin. Epidemiol. **123**, 177–179 (2020). https://doi.org/10.1016/j.jclinepi.2020.03.016
14. Munn, Z., Peters, M.D., Stern, C., Tufanaru, C., McArthur, A., Aromataris, E.: Systematic review or scoping review? guidance for authors when choosing between a systematic or scoping review approach. BMC Med. Res. Methodol. **18**, 1–7 (2018)
15. NATO Standardization Office (NSO). NATO standard app-6, NATO joint military symbology. Standard Edition D, Version 1, North Atlantic Treaty Organization (NATO) (2017)
16. Pistono, F., Yampolskiy, R.V.: Unethical research: how to create a malevolent artificial intelligence. In: Proceedings of Ethics for Artificial Intelligence Workshop (AI-Ethics-2016), pp. 1–7 (2016)
17. Renaud, K., Warkentin, M., Westerman, G.: From ChatGPT to HackGPT: meeting the cybersecurity threat of generative AI. MIT Sloan Management Review (2023). Reprint #64428

18. Smith, J., Theisen, C., Barik, T.: A case study of software security red teams at Microsoft. In: 2020 IEEE Symposium on Visual Languages and Human-Centric Computing (VL/HCC), pp. 1–10. IEEE (2020). https://doi.org/10.1109/VL/HCC50065.2020.9127203

19. Sommestad, T., Hallberg, J.: Cyber security exercises and competitions as a platform for cyber security experiments. In: Jøsang, A., Carlsson, B. (eds.) Secure IT Systems. LNCS, vol. 7617, pp. 47–60. Springer, Heidelberg (2012). https://doi.org/10.1007/978-3-642-34210-3_4

20. Truong, T.C., Diep, Q.B., Zelinka, I.: Artificial intelligence in the cyber domain: offense and defense. Symmetry **12**(3), 410 (2020)

21. Wang, C., Chen, J., Yang, Y., Ma, X., Liu, J.: Poisoning attacks and countermeasures in intelligent networks: status quo and prospects. Digit. Commun. Netw. **8**(2), 225–234 (2022)

22. Wang, W., Siau, K.: Artificial intelligence, machine learning, automation, robotics, future of work and future of humanity: a review and research agenda. J. Datab. Manag. **30**(1), 61–79 (2019). https://doi.org/10.4018/JDM.2019010104

23. Ward, D., Wooderson, P.: Automotive Cybersecurity: An Introduction to ISO/SAE 21434, p. 106. SAE International (2021)

24. Wilhelmson, N., Svensson, T.: Handbook for planning, running and evaluating information technology and cyber security exercises. In: The Swedish National Defence College, Center for Asymmetric Threats Studies (CATS) (2014)

25. Yamin, M.M., Ullah, M., Ullah, H., Katt, B.: Weaponized AI for cyber attacks. J. Inf. Secur. Appl. **57**, 102722 (2021)

26. Yuen, J.: Automated Cyber Red Teaming. DSTO Defence Science and Technology Organisation, Edinburgh (2015)

27. Zhou, W.C., Sun, S.L.: Red Teaming Strategy: Huawei's Organizational Learning and Resilience, pp. 299–317. Springer, Cham (2020). https://doi.org/10.1007/978-3-030-47579-6_13

28. Zouave, E., Bruce, M., Colde, K., Jaitner, M., Rodhe, I., Gustafsson, T.: Artificially intelligent cyberattacks. Tech. Rep. FOI, Swedish Defence Research Agency, FOI (2020)

Third-Party Data Leaks in the Websites of Finnish Social and Healthcare Districts

Panu Puhtila, Esko Vuorinen, and Sampsa Rauti[(⊠)]

University of Turku, 20014 Turku, Finland
{papuht,etvuor,sjprau}@utu.fi

Abstract. With digitalization, the use of essential social and healthcare services online has become increasingly prevalent. In this paper, we conduct a survey on the websites of Finnish social and healthcare districts and determine to what extent, if any, they leak their users' personal data to third parties through the use of the collection and tracking of user data and actions with the web analytics tools. Our findings show that 82.6% of the studied websites leaked personal data to outside actors, but the extent and contents of these data leaks varied. Our study also demonstrates that in many cases, privacy policies of the studied websites do not always report personal data items transferred to third parties and fail to adequately inform users. The cookie banners of the studied websites were also found to contain several dark patterns.

Keywords: social and health websites · data leaks · data concerning health · online privacy · third-party services · SOTE

1 Introduction

The past decade has seen the proliferation of several technologies that have brought the internet to the large segments of population, ushering in an unprecedented increase in the use of online services in day-to-day activities. As technology advances, user data, especially personally identifiable information, has emerged as a valuable resource collected and monetized through targeted marketing on websites. In practice, this data collection is enabled by web analytics tools, a class of applications that monitor and survey the website user actions. In principle, these tools are designed to help the website proprietors to better manage their services. However, at the same time these tools often leak the data they gather outside the websites they are deployed in, usually to servers run and owned by the same corporations that have created these analytics tools in the first place. Laws have been enacted to prevent third-party data leaks without the user's consent and strengthen personal privacy. All too frequently, however, enforcement is lacking, and compliance with privacy regulations falls short of addressing the magnitude of the issue they intend to rectify.

This study is part of our research on third-party data leaks on Finnish websites, conducted as part of IDA (Intimacy in Data-Driven Culture) research

© The Author(s), under exclusive license to Springer Nature Switzerland AG 2024
A. Rocha et al. (Eds.): WorldCIST 2024, LNNS 985, pp. 139–152, 2024.
https://doi.org/10.1007/978-3-031-60215-3_14

project. In this paper, we study the data leaks taking place in the Finnish public sector social and healthcare district websites. As an example case, we study how their online services for the alcohol- and drug-dependent people leak personal information to third parties. We will also review the privacy policies and cookie banners of these websites and determine whether their contents are in accordance with the actual data collection happening at the website. Due to the delicate nature of health related data, it is very important that no identifying information should end in wrong hands. For example, simply the knowledge that someone has sought help because of a medical condition can be extremely harmful and stigmatizing, potentially leading to severe problems in social and professional circles.

Social and healthcare services provided by the Finnish public sector have undergone a considerable change in the past few years as the result of the nationwide project called Social Welfare And Healthcare reform [9,10,13,17], during which the organization, production and government of these services was transferred to a new jurisdictional body, the newly instituted wellbeing services counties. In Finnish this reform was termed "SOTE-uudistus", and due to this we have chosen to use the term "SOTE-portals" when discussing the websites inspected in this study. SOTE is a Finnish acronym of "sosiaali- ja terveys", which directly translates to "social and health". This reorganization was completed in a tight schedule, which may be a factor in explaining our results, namely, the presence of several third-party data leaks on the websites we studied. Streamlining the production of the web services at the cost of decreased protection of user privacy may have contributed to the data leaks discovered in our study. To the best of our knowledge, the current study is the first study on the privacy of the new websites of Finnish social and healthcare districts.

The rest of the paper is organized as follows. In Sect. 2, we take a look at the previous research conducted on similar subjects. In Sect. 3, the methodology and the setting for our research is laid bare. In Sect. 4, the results of our research are presented. In Sect. 5, we discuss the implications that can be drawn from our results. Finally, in Sect. 6 we present the conclusions of the study.

2   Previous Research

The situation of the Finnish social and healthcare reform is quite unique, and research concentrating on exactly this kind of phenomenon, that is, potential data leaks in social and healthcare districts' websites after such a reform, has not been studied in previous research. However, there is an ample body of research done on the data leaks and privacy violations in medical websites in general, as well as in a more general context of social and healthcare websites. To position our study within the context of prior research, we present some of this previous work here.

As far back as 2012, Masters [14] studied the data collection on the websites of the National Medical Association members in the USA, and came to the conclusion that even then, 47% of the studied websites gathered user data. Huesch studied the privacy threats in medical websites in 2013 [11], and concluded that slightly over one third of the inspected websites leaked their user information to third parties. In the same year, Brown and Levy [2] published a paper which detailed a proof-of-concept design for a tool to benchmark the information collection practices of pharmaceutical websites. During the same period, Burkell and Fortier [3, 4] conducted a study which showed how medical websites did not correctly disclose their data collection activities, leading to users falsely giving consent to such practices. They also detailed how consumer health websites constantly collected personal data on their users with analytics tools to intentionally build detailed profiles of them.

More recently, Surani et al. [16] came to a conclusion that both the websites and applications used in the mental health services often leak data and exhibit also other kinds of privacy and security risks. Zheutlin et al. [20] conducted a research on how USA-based government, non-profit and commercial health-related websites collected the user data. Friedman et al. [6] show in their recent study in which ways the tracking applications in hospital websites threaten the user privacy, also putting hospitals in a legally questionable situation.

Yu et al. [19] studied a similar phenomenon, by conducting a wide-scale automated scan on tens of thousands of hospital websites all over the world, which revealed that 53.5% of them used analytics tools that collected the data of their website users. Friedman et al. researched in 2022 [5] the use of web analytics tools by abortion clinics, and came to a conclusion that the vast majority of them (99.1%) deployed at least one analytics service which leaked the user data to third-party actors. Huo et al. [12] found in their research on the patient web portal privacy that 14% of them leaked sensitive data such as names and phone numbers outside the website domain. Schnell and Roy published a paper in 2022 [15] inspecting the hospital website design, in which they concluded that the design actively inhibited the user from finding the privacy policy, thus making the users unable to consent to the data collection practices. Wesselkamp et al. [18] developed a browser extension in 2021 to detect third-party tracking cookies used on websites. Then they used it to research 385 medical websites operating in the EU area, and whether they collected the user data. The researchers discovered that 62% of the inspected websites used web analytics tools automatically, before any consent to data collection was expressed, and 15% collected the data even if the consent was not given.

Previous research appears to suggest there is a strong correlation between the use of web analytics tools in websites and data leakages to third parties, often resulting in severe breaches of personal privacy. Further investigation is essential to comprehensively understand and address this issue. Our study offers an in-depth examination of data leaks in the websites of Finnish social and healthcare districts, delving into the nature of personal data leaked to third parties.

3   Methodology and Study Setting

In this study, we inspected 23 Finnish social and healthcare district websites (referred as SOTE-portals from this point onwards). The websites chosen to be studied correspond to the official social and healthcare districts, which were instituted during the Social Welfare and Healthcare reform described previously. The list of these websites is also available online[1].

The data leakages in websites were studied in the following manner. First, a researcher navigated to the website. After this, Google Chrome Developer Tools (referred to as devtools from this point onwards) were turned on, and all caches were cleared and disabled. All cookies were consented to when arriving at the website. Then the website was refreshed to ensure that no cached information would distort the test results.

The testing sequence varied between different websites due to their different designs and architectures, but the general pattern was always the same: First the researcher entered a Finnish expression for drug- and alcohol-dependence "päihdeongelmat" ("problems with intoxicants") into the search console of the website in question. Then the researcher clicked the link that seemed the most relevant in finding help to alcohol- or drug-dependency from those listed by the search functionality. If clicking this link led to subsequent links, the most relevant option for attaining help was always followed, until the link trail either came to an end or led outside the studied website.

During the procedure described above, all network traffic was recorded with the devtools and saved as HAR-files[2] for later analysis. The HAR-files were further filtered to find only the instances where HTTP requests were made outside the domain of the social and healthcare provider, in other words, to third parties. We specifically concentrated on three chief factors:

1. Whether the URL address of the visited page was leaked.
2. Whether the clicking of the link to a page about getting help for addiction was leaked.
3. Whether the used search term was leaked.

The privacy policies and cookie consent banners were read and saved either as .jpg or .pdf format for later inspection. In studying the privacy policies we paid specific attention to four different factors: Whether all third parties were named in the document, and whether the three factors discussed previously (URL leaks, link click leaks, search term leaks) were mentioned at all in the privacy policy. In addition to this, we also examined whether the cookie consent banners used in the SOTE-portals exhibited design choices which can be interpreted as dark patterns. In doing this, we used the dark pattern categorizations formulated by the European Cookie Banner Taskforce[3], from which we chose four categories to examine:

1. Absence of "decline cookies" button on the first layer of the cookie banner

---

[1] https://soteuudistus.fi/hyvinvointialuekartta.

[2] The HTTP Archive format is a file format for recording a web browser's interactions with a website.

[3] https://edpb.europa.eu/our-work-tools/our-documents/other/report-work-undertaken-cookie-banner-taskforce_en.

2. Pre-ticked consent boxes
3. Deceptive use of colors
4. Deceptive use of contrasts

It should be noted that in determining the geographical location of the servers where the leaked information was sent to we have used the `iplocation.net` service, which combines the information from eight different IP-locator services. Because these geographical locations are difficult to pinpoint in the era of cloud services, we have accepted the location to be included in our results only if all eight sources indicated the same destination.

Finally, as this paper is about leaks of personal data, we present a definition for this concept. The one given in the GDPR of the European Union and also used by the Finnish Office of the Data Protection Ombudsman is sufficient for the purposes of our study[4]. Thus, personal data is defined as "all data related to an identified or identifiable person". By this definition, technical information such as IP addresses, device identifiers, accurate location data or any data point that identifies the user of the website counts as personal data. It must also be noted that while many technical details such as device type or screen resolution alone are not sufficient to identify someone, together these data items can be used for assisting in identification of a specific person. Hence, they can be considered as personal data.

4 Results

We found that 21 out of 23 (82.6%) of the studied websites leaked personal data to third parties. A slight majority of the inspected websites, 12 out of 23 (52.17%), leaked data only to one third-party actor. Moreover, 5 out of 23 (21.7%) leaked data to two actors, 2 out of 23 (8.6%) to 3 actors and only one (4.3%) studied website leaked data to 4 different third parties. This comprises the total of 32 leaky data collection tools between 23 websites. Two of the studied websites did not leak data at all outside their own domains. Neither of these two websites had any analytics tools in use, which indicates a clear connection between the leaking of user data and the use of web analytics tools in the first place.

In total, 11 different third-party services were found to be used between the 23 websites we inspected. The most commonly encountered third-party service and analytics tool among the studied websites was Google Analytics. This result does not come as a surprise since Google Analytics is the largest "free" analytics tool in circulation today, and has previously [1,8] been shown to take the top spot as the most used analytics tool in many studies. Google Analytics was used in 9 out of 23 (39.1%) of the websites we studied in this research.

Behind Google Analytics came the services Reactandshare and Siteimprove-analytics, which were both found in 5 out of 23 websites (21.7%). Third place in popularity was taken by Istekkipalvelut.fi, which was used in 4 out of 23 websites

---

[4] https://gdpr-info.eu/.

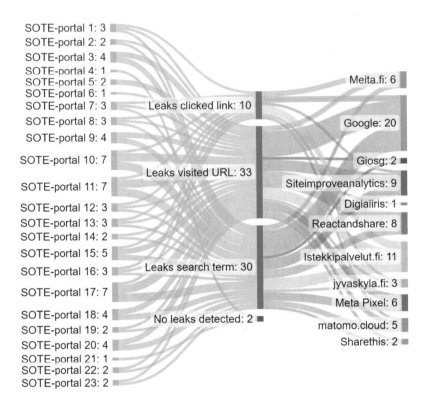

Fig. 1. Total numbers of data leaks in the SOTE-portals and the third parties the data is leaked to.

(17.4%). Curiously, a large and often encountered website analytics tool, Meta Pixel, was quite rarely used by the websites inspected here, and was present in only 3 out of 23 (13.0%) instances. Of the remaining seven third-party tools, Meita.fi and Matomo were both encountered in two different domains (8.6%). It is important to understand, however, that Meita.fi provides a service that itself deploys Matomo as the tool to collect data. The difference here is that two of the websites had Matomo deployed locally at their own clouds, while two used the larger service package from Meita which includes Matomo analytics. Each of the remaining 5 third parties was encountered only once.

It is very important to note that although Istekkipalvelut and Meita are listed as third parties here because of their own domains, they are actually in-house companies owned by public sector bodies. Therefore, data transfers to these parties are not third-party data leaks in the usual sense of the term but rather intended functionality. The same goes for jyvaskyla.fi, which is a domain of a Finnish city. Therefore, these data transfers are very different from data leaks to companies such as Google and Meta in terms of privacy risks to the user. Of course, it can still be questioned whether certain delicate page visits or

Table 1. Data leaks in the social and healthcare websites.

Website	Clicked link leaks	Visited page leaks	Search term leaks
SOTE-portal 1	X	X	X
SOTE-portal 2		X	X
SOTE-portal 3		X	X
SOTE-portal 4			
SOTE-portal 5		X	
SOTE-portal 6		X	
SOTE-portal 7	X	X	X
SOTE-portal 8	X	X	X
SOTE-portal 9		X	X
SOTE-portal 10	X	X	X
SOTE-portal 11	X	X	X
SOTE-portal 12	X	X	X
SOTE-portal 13	X	X	X
SOTE-portal 14		X	X
SOTE-portal 15	X	X	X
SOTE-portal 16	X	X	X
SOTE-portal 17	X	X	X
SOTE-portal 18		X	X
SOTE-portal 19		X	X
SOTE-portal 20		X	X
SOTE-portal 21			
SOTE-portal 22		X	X
SOTE-portal 23		X	X

link clicks, for instance, should be ever stored at all, even if the data is hosted by a trustworthy party.

The detected data transfers can be seen in Fig. 1. In the figure, the numbers found besides the names of data collection tools and SOTE-portals represent the total number of all leakages in the three categories we examined. We can see that Google clearly receives the largest number of data transfers.

The most commonly leaked item of interest to our research was the URL of the visited page, which was, apart from the two websites which did not leak any data, leaked by 21 of the studied websites to at least one third-party actor. The second next leaked item was the search term, which was leaked in 19 out of 23 (86.2%) websites. The information about the clicking of the link that leads to a "seeking for help" page was leaked in 10 out of 23 instances (43.5%). These findings are unacceptable, considering that we are talking about online resources meant for people who may be suffering from very stigmatizing personal problems.

Table 2. Discrepancies between the studied privacy policies and actual transmitted data.

Website	Third parties mentioned	Search term	Visited page	Clicked link
SOTE-portal 1		X	X	X
SOTE-portal 2		X	X	
SOTE-portal 3		X		
SOTE-portal 4				
SOTE-portal 5	X		X	
SOTE-portal 6	X		X	
SOTE-portal 7		X	X	X
SOTE-portal 8	X	X	X	X
SOTE-portal 9		X		
SOTE-portal 10		X	X	X
SOTE-portal 11		X		X
SOTE-portal 12		X	X	X
SOTE-portal 13		X		X
SOTE-portal 14	X	X	X	
SOTE-portal 15		X		X
SOTE-portal 16		X		X
SOTE-portal 17	X	X	X	X
SOTE-portal 18		X		
SOTE-portal 19	X	X		
SOTE-portal 20		X		
SOTE-portal 21				
SOTE-portal 22		X	X	
SOTE-portal 23		X	X	

These results are presented in Table 1. In the table, blue means that no leakages occurred, and red that they did.

Along with the data items we specifically focused on in this research, the studied websites also leaked identifying information. IP addresses and User-Agent strings, leaked with HTTP requests by default, are such data items. For user identification, Google Analytics also uses the cid number, a variable that is assigned to every unique browser-device pair. There are many other technical details that can be used as parts of the digital fingerprint for an individual user. For instance, many of the studied third parties (Google, Giosg, Siteimprovanalytics, Digiaiiris, Meta Pixel, jyvaskyla.fi and Istekkipalvelut.fi) also received the screen size of the used device, which can help in approximating the identity of the used device, and consequently, the user of the said device.

Our survey of the privacy policies found similar deficiencies as were present in the actual data collection, the details of which can be seen in Table 2. The

Table 3. Dark patterns found in the cookie consent banners of the SOTE-portals.

Website	Asks for consent consent	Reject cookies button in first layer	Pre-ticked consent boxes	Deceptive use of colors	Deceptive use of contrasts
SOTE-portal 1		X		X	X
SOTE-portal 2				X	X
SOTE-portal 3				X	X
SOTE-portal 4					
SOTE-portal 5					
SOTE-portal 6					
SOTE-portal 7				X	X
SOTE-portal 8				X	X
SOTE-portal 9				X	X
SOTE-portal 10				X	X
SOTE-portal 11				X	X
SOTE-portal 12				X	X
SOTE-portal 13				X	X
SOTE-portal 14					
SOTE-portal 15					
SOTE-portal 16				X	X
SOTE-portal 17					
SOTE-portal 18				X	X
SOTE-portal 19				X	X
SOTE-portal 20			X	X	X
SOTE-portal 21					
SOTE-portal 22					
SOTE-portal 23				X	X

black boxes mean that the data collection of the specific type did not happen at the website and was not applicable to the privacy policy. The blue color means that the privacy policy adequately described this form of data collection, and red that it did not, in the instances where data collection happened. While 15 out of 23 websites did mention all of the third parties in their privacy policies, none of the inspected websites informed the user that the information about clicking of the links or the search terms used would be collected. This result is quite frankly astounding, in a negative way. As it was so completely uniform across these different websites, it may be that the people responsible for penning down the privacy policies have not understood that these two data items can be leaked or could be related to sensitive data leaks in the first place. On the other hand, one could argue that from a legal viewpoint that mentioning the collection of just the visited URL covers at least the search term as well, since it was always leaked as a part of the URL. However, an average user can not be expected to understand that leaking an URL address may also mean that the search term leaks.

Moreover, only 9 out of 21 (39.1%) privacy policies mentioned that the visited URL would be collected, which can be considered a severe problem, as all of the sites which had any kinds of data collection did leak this piece of information.

Two of the studied websites did not have any kind of privacy policy document at all. One of these also did not leak any kind of information to third parties and deployed no detectable tracking measures, which can be seen as a mitigating factor in the lack of privacy policy. The other one did leak data, however, and the lack of privacy policy or it being inaccessible to the user of the website is a direct breach of the GDPR, and it can lead to legal consequences for the parties involved. All in all, the contents of the privacy policies we encountered were inadequate, and can not be considered to give enough information to the users of the website so that they could make an informed decision about whether to consent to data collection.

Table 3 illustrates our findings in regards to dark patterns used in the design of the cookie banners. As can be seen here, the majority of the SOTE-portals asked for consent to data collection, used deceptive colors and contrasts in their cookie banners and otherwise did not use design practices that could be interpreted as dark patterns. Only one of the studied SOTE-portals lacked the "Reject cookies" button in the first layer of the cookie consent banner, and only one used pre-ticked consent boxes in their banner. Both of these attributes are considered by the Cookie Banner Taskforce to annul the consent, as the user can not be considered to be making an informed decision in either case. Five of the studied SOTE-portals did not exhibit any dark patterns in their cookie consent banner designs. Three of the studied websites—marked in black in the table—did not have any kind of cookie consent banner at all. Also, it should be noted that the singular black boxes in the column "Pre-ticked consent boxes" mean that the website in question did not present the user any option on what types of cookies to consent to, but rather just a general "Accept/Decline cookies" option. SOTE-portal number 4 had a perfect cookie consent banner without any dark patterns, but at the same time did not use any data collection tools. It is the only portal that gets a perfect score in all of our tests.

The destinations of the leaked data, as far as we could determine, were mostly within the jurisdiction of the European Union. In 26 instances of leaky data collection tools where we could be sufficiently confident in our ability to determine the destination of the data, only once did the leaked data end up in servers outside the EU area. In this one instance, the servers were most likely located in the USA. Most common of the destinations for the data were in Finland in 10 of the inspected cases, followed by Ireland (6 instances), Sweden (5 instances) and Germany (4 instances). These results can be considered to be very good in terms of keeping the collected data within the borders of the EU.

What is noteworthy in our results was the amount of domestic data analytics providers versus global corporations we encountered during this research, as 6 out of 11 encountered data collection tools were sourced from Finnish IT companies. Two of these are well-known operators in the field of providing several kinds of information technology services for the needs of the Finnish public sector (Istekkipalvelut and Meita), and one was a service operated by a Finnish city, Jyväskylä. In the case of two websites the data collection was facilitated by Matomo which was deployed by the proprietors of these services themselves to a private cloud. The large percentage of domestic services is likely to be linked to the nature of these websites as part of the Finnish public healthcare and social service infrastructure.

In conclusion, the results we obtained can be considered concerning, but also moderately hopeful. They are alarming because it became apparent that 21 out of 23 (82.6%) of the studied websites leaked somewhat sensitive information to third parties. Every single website which had some kind of analytics tool leaked data to at least one third party. Yet, at the same time this result was hopeful, as 12 out of 23 (52.17%) of the studied SOTE-portals used no more than one tracking tool for data collection. Compared to the results of previous studies [8], which have also targeted the public sector social and healthcare related websites in Finland, this is quite promising. In the contemporary internet ecosystem where the use of web analytics tools is more the norm than an exception, and having a plethora of such third-party services deployed at every website is very common, over half of the SOTE-portals having only one such tool deployed can be considered a moderately good result in regards of user privacy. Also, the fact that the majority of the servers where the leaked data ended up were located in the EU is positive. Of course, in the cases of actors like Google and Meta, data may be transferred beyond European borders and jurisdictions, even when initially stored on EU servers.

5   Discussion

Our findings reveal that the transmission of sensitive information, including visited pages and search terms, to third parties is more common than not leaking any data. Furthermore, nearly half of the websites leaked information about users' intent to seek help, often through link clicks. These leaks are partially facilitated by unclear privacy policies that fail to adequately inform users and the presence of various dark patterns in cookie banners.

Considering the results we have obtained in this study, it is important to pose the question of whether there is any real reason for using analytics tools as parts of medical websites, whether operated by the public sector or by the private enterprise. It is extremely doubtful whether there is enough added value gained from using these tools to justify their use. The web analytics are meant, in the ideal situation, to help the website proprietors to identify how their users interact with the website. Designing web-based healthcare services in the most user-friendly ways can be achieved through means other than invasion of the privacy of the users such as usability testing. Considering the sensitive data the studied web services process, it should be obvious that such personal data should never fall into wrong hands outside the health or social service the user is interacting with. While it can be argued that the risk of abusing the data is not necessarily very large, it exists nonetheless. As this kind of exposure can lead to serious consequences for an individual's personal and professional lives, even a small risk is unacceptable.

On the bright side, it can be argued that both the relatively small amount of encountered data collection tools, and on the other hand the fact that the majority of the leaked data did not end up in servers outside the area of EU, are very good results. It is, after all, very common in leaks like this that the leaked data ends up overseas, often into servers operated in the USA. It is also quite common for the website operators to deploy many different analytics tools, which all have more or less similar data collection profiles, thus worsening the situation with leaks to numerous actors and locations.

Developers and maintainers of the SOTE-portals should ensure the user privacy by conducting a careful review of the data collection tools in use. If such surveillance tools are deemed absolutely necessary for the operation of the website in question the workings of these applications should be configured so that no leakages occur, by conducting a thorough network traffic analysis akin to what we have done in this study. Doing this demands neither special expertise, special software nor large amounts of time. In essential and critical services like SOTE-portals, there are no good justifications to not perform such testing. If a specific analytics tool cannot be configured in such a way that it would not leak the user data outside the domain it is deployed in it should be abandoned, as there are other options for web analytics that can be used without any data leaking to third parties, such as Matomo [7].

6   Conclusions

In this paper, we have studied the Finnish SOTE-portals from the perspective of user privacy, and our findings reveal clear needs of improvement. Apart from the two websites that did not use any analytics tools at all, all websites that used web analytics also transmitted sensitive personal data on their users to third parties. The SOTE-portals under examination exhibited several data leaks, including URL addresses of visited pages, search terms, and clicks on links to help-seeking pages, combined with identifying information such as IP addresses. These leaks can potentially expose sensitive personal data, such as an individual's intent to seek help for substance abuse. Since the websites examined cater to individuals dealing with health issues that can affect not only their physical well-being but also their social standing, data leaks are especially serious in this context.

Our findings indicate that severe oversight and negligence on the issue of user privacy has taken place when implementing the SOTE-portals. This situation should be remedied post-haste, as longer the current state persists, more likely it is to become entrenched as the "new normal". To fix the issues we discovered in the SOTE-portals demands that the proprietors conduct a thorough assessment of the data collection technologies they use, and discard those which can not be configured properly. Improving the privacy of social and healthcare services is not just a matter of the moral issue of exposing the users of these websites to a risk of being identified, but also legal considerations in regards to the requirements set in the GDPR of the European Union.

Acknowledgements. This research has been funded by Academy of Finland project 327397, IDA—Intimacy in Data-Driven Culture.

References

1. Bailey, J., Laakso, M., Nyman, L.: Look who's tracking: an analysis of the 500 websites most-visited by finnish web users. Informaatiotutkimus **38**(3–4), 20–44 (2019)
2. Brown, S.D., Levy, Y.: Towards a development of an index to measure pharmaceutical companies' online privacy practices. Online J. Appl. Knowl. Manag. **1**(1), 93–108 (2013)
3. Burkell, J., Fortier, A.: Consumer health websites and behavioural tracking. In: Proceedings of the Annual Conference of CAIS/Actes du congrès annuel de l'ACSI (2012)
4. Burkell, J., Fortier, A.: Privacy policy disclosures of behavioural tracking on consumer health websites. In: Proceedings of the American Society for Information Science and Technology, vol. 50, pp. 1–9. Wiley Online Library (2013)
5. Friedman, A.B., Bauer, L., Gonzales, R., McCoy, M.S.: Prevalence of third-party tracking on abortion clinic web pages. JAMA Intern. Med. **182**(11), 1221–1222 (2022)
6. Friedman, A.B., et al.: Widespread third-party tracking on hospital websites poses privacy risks for patients and legal liability for hospitals. Health Aff. **42**(4), 508–515 (2023)
7. Gamalielsson, J., et al.: Towards open government through open source software for web analytics: the case of matomo. JeDEM-eJ. eDemocracy Open Gov. **13**(2), 133–153 (2021)
8. Heino, T., Carlsson, R., Rauti, S., Leppänen, V.: Assessing discrepancies between network traffic and privacy policies of public sector web services. In: Proceedings of the 17th International Conference on Availability, Reliability and Security, pp. 1–6 (2022)
9. Hiilamo, H.: Why did social and healthcare services reform fail in finland? Socialmedicinsk tidskrift **97**(3), 433–441 (2020)
10. Hirvensalo, E., Asko-Seljavaara, S., Haahtela, T., Leppäniemi, A., Tukiainen, E.: Sote-uudistus ei toteuta säästöjä eikä parempaa hoitoa. Suomen lääkärilehti (2017)
11. Huesch, M.D.: Privacy threats when seeking online health information. JAMA Intern. Med. **173**(19), 1838–1840 (2013)
12. Huo, M., Bland, M., Levchenko, K.: All eyes on me: inside third party trackers' exfiltration of phi from healthcare providers' online systems. In: Proceedings of the 21st Workshop on Privacy in the Electronic Society (WPES 2022), pp. 197–211. Association for Computing Machinery, New York (2022)
13. Jalonen, H.: Sote-uudistus: mitä, kuka, missä ja miten? (2021)
14. Masters, K.: The gathering of user data by national medical association websites. Internet J. Med. Inform. **6**(2) (2012)
15. Schnell, K., Kaushik, R.: Hunting for the Privacy Policy - Hospital Website Design (2022)
16. Surani, A., et al.: Security and privacy of digital mental health: an analysis of web services and mobile apps. In: Conference on Data and Applications Security and Privacy (2023)
17. Vauramo, E.: Miten sote-uudistus toteutetaan?
18. Wesselkamp, V., Fouad, I., Santos, C., Boussad, Y., Bielova, N., Legout, A.: In-depth technical and legal analysis of tracking on health related websites with ernie extension. In: Proceedings of the 20th Workshop on Workshop on Privacy in the Electronic Society (WPES 2021), pp. 151–166. Association for Computing Machinery, New York (2021)

19. Yu, X., Samarasinghe, N., Mannan, M., Youssef, A.: Got sick and tracked: privacy analysis of hospital websites. In: 2022 IEEE European Symposium on Security and Privacy Workshops (EuroS&PW), pp. 278–286. IEEE (2022)
20. Zheutlin, A.R., Niforatos, J.D., Sussman, J.B.: Data-tracking on government, non-profit, and commercial health-related websites. J. Gen. Internal Med. 1–3 (2021)

A Model for Assessing the Adherence of E-Identity Solutions to Self-Sovereign Identity

Cristian Lepore[1]([⊠]), Romain Laborde[1], Jessica Eynard[2], Mohamed Ali Kandi[1], Giorgia Macilotti[3], Afonso Ferreira[3], and Michelle Sibilla[1]

[1] IRIT, 118 Route de Narbonne, 31400 Toulouse, France
cristian.lepore@irit.fr
[2] Université Toulouse Capitole, 2 rue du Doyen-Gabriel-Marty, 31042 Toulouse, France
[3] CNRS, 16 Av. Edouard Belin, 31400 Toulouse, France

Abstract. Self-Sovereign Identity (SSI) represents a new concept to manage digital identities, aiming to empower individuals by giving them control over their data. However, the concept is still elusive, and many design patterns coexist without an agreed-upon standard which allows anyone to build identity systems while declaring adherence to SSI. We contribute by formalizing a definition of Self-Sovereign Identity and a corresponding evaluation model of digital identity solutions. We then demonstrate our model value with an in-depth analysis of VIDchain, a business product that promotes Self-Sovereign Identity services in compliance with European regulations. Ultimately, our analysis discusses the quest for a perfect SSI solution and supplies end users with a tool to choose the SSI e-identity solution that best fits their needs.

Keywords: Assessment · Digital Identity · eIDAS Bridge · Self-Sovereign Identity Principles · Validated ID · VIDchain

1 Introduction

Digital identity is an essential factor in economic growth for businesses and governments. In 2022, the global e-identity verification market was worth $27.9 billion, with an expected growth of over 16% to 2030 [1]. However, every time an App or website asks us to create an e-identity, we have no idea what happens to our data. Today, nearly 72% of users wish to have more control over their e-identities.[1] This is where Self-Sovereign Identity comes in.

Self-Sovereign Identity (SSI) is a relatively new approach to manage e-identities that aims to give end users control of their identity information. A definition of SSI is still elusive [2], but reference architectures have been proposed. The reference architecture consists of an issuer providing verifiable credentials (VCs),

---

[1] Digital Identity for all Europeans; a personal digital wallet for EU citizens and residents. https://commission.europa.eu/strategy-and-policy/priorities-2019-2024/europe-fit-digital-age/european-digital-identity_en.

© The Author(s), under exclusive license to Springer Nature Switzerland AG 2024
Á. Rocha et al. (Eds.): WorldCIST 2024, LNNS 985, pp. 153–163, 2024.
https://doi.org/10.1007/978-3-031-60215-3_15

which include specific claims about the subject/holder. The subject may differ from the holder. The subject/holder composes a verifiable presentation (VP) by combining VCs and delivers it to a verifier. Ultimately, the subject can reveal information to verify their identity to the requester without disclosing it to others. Thus, the subject is sovereign in the use of their credentials.

A legal trust framework is crucial for e-identities to be recognized under the country's jurisdiction. In Europe, the eIDAS regulation supplies a legal framework for the governance of e-identities [3]. Despite the recent amendment that aims to give more control to end users, the regulation is not a governance framework for SSI. Thus, the problem of issuing legal identities in an "SSI environment" is relegated to ad-hoc projects, for example, the eIDAS Bridge. Through the eIDAS bridge, Validated ID – a pioneering company in providing Self-Sovereign Identity solutions – bridges the gap between SSI and legal identity utilizing the development of VIDchain [4].

Today, the emergence of Self-Sovereign Identity solutions outlines the importance of assessing the adherence of those solutions to SSI. Previous studies stress the significance of defining specific criteria for evaluation; otherwise, the contribution of their content analysis is limited [5,6]. Therefore, a rigorous definition of Self-Sovereign Identity would facilitate the design and validation of solutions, addressing their completeness and correctness [7]. Thus, we pose the following research questions:

RQ1: What are the principles of Self-Sovereign Identity? We outline the concept of Self-Sovereign Identity through a rigorous definition of principles that considers the interdisciplinary of the subject. We systematize the literature by analyzing articles on ACM, ArXiv, Google Scholar, and IEEE Xplore. We outline concepts, relationships, and rules governing identity ecosystems' entities and provide a formal specification of principles of SSI.

RQ2: Can we assess any SSI system based on those principles? We delineate a model based on our implementable tweak of SSI principles to assess any worldwide digital identity system. We then demonstrate our model value with an in-depth analysis of VIDchain, a product designed to issue eIDAS-compliant Self-Sovereign Identities. We fill the gap between SSI theory and practical design. Our findings allow us to propose a more pragmatic definition of SSI based on the overall performance of the e-identity system. In the long run, we aim to enable future startups and governments to rank solutions, spot weaknesses, and intervene accordingly.

The remainder of this paper continues as follows. The following Section describes our method to structure the research field of SSI. Section 3 provides our contribution as 1) a holistic definition of SSI and 2) a model to assess e-identity solutions. Section 4 demonstrates our model value through a comprehensive analysis of VIDchain. We discuss our findings in Sect. 5 and limitations in Sect. 6 before concluding with avenues for future research.

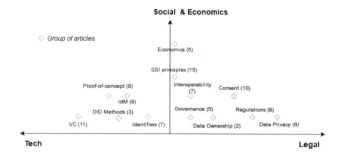

Fig. 1. A systematization of knowledge in a two-axes chart.

2 Methodology

The objective is to structure the research field of Self-Sovereign Identity to spot literature gaps and build new constructs [8]. A systematic review study provides a coarse-grained overview of the research field through several steps as follows [9].

1. *Defining research questions.* We produced research questions *RQ1* and *RQ2* as outlined in the introduction. From keywording the research questions, we provided the following search strings. We shuffled keywords for better output and corrected strings to avoid wildcards (e.g., SSI and Self-Sovereign Identity and assessing) [10].
   – RQ1: "SSI principles"
   – RQ2: "Assess" AND "SSI system" AND "SSI principles"
2. *Searching.* We used search strings to hit articles with relevant keywords in titles and abstracts from ACM, ArXiv, and IEEE Xplore and meta-search engines of academic sources like Google Scholar. That yielded 250 results.
3. *Paper Screening.* We screened abstracts and conclusions to filter out non-pertinent results through inclusion/exclusion criteria based on the subject matter of interest (SSI), publication year, originality of the work, and proofs-of-concept. We also excluded duplicate results, collecting 95 articles subject to full review.
4. *Classification scheme.* We read in full the 95 articles and pencilled out information about their objectives, outcomes, state of knowledge, computational method, worked part, and limitations. We rendered thirteen groups based on this information, from Consent management, identifiers, DID Methods, Data ownership/control, Data privacy, Economics, Governance, Identity Models (IdM), Interoperability challenges, Proof-of-concept, Regulations, SSI principles, Verifiable Credentials. We then assigned articles to the most relevant group.
5. *Data extraction.* We plotted those groups in the two-axis chart of Fig. 1, with the horizontal axis (x-axis) sketching technical and legal matters (left to right) and a vertical axis (y-axis) grading the social & economic aspects. Diamonds

Table 1. The summary of principles and taxonomies of SSI by different authors.

Principles	Sovrin(2016) [16]	Andrieu(2016) [11]	Ferdous(2019) [12]	Gilani(2020) [13]	Sheldrake(2019) [14]	BkThDVr(2022) [15]
Existence	Control	Control	Foundational	Foundational	Foundational	Personal Data
Control	Control	Control	Control	Foundational	Foundational	Control
Access	Portability	Acceptance	Foundational	Foundational	Foundational	Control
Transparency	Portability	Acceptance	Sustainable	Sustainable		Usability
Persistence	SecurityControl	Control	Security	Security	Foundational	Personal Data
Portability	Portability	AcceptanceZero-Cost	Flexibility	Flexibility		Personal Data
Interoperability	Portability	Acceptance	Flexibility	Flexibility		Usability
Consent	Control	Control	Control	Foundational		Personal Data
Protection	Security	Acceptance	Security	Security		Personal Data
Minimization	Security	Control	ControlFlexibility	FoundationalFlexibility		Personal Data
Autonomy			Foundational			
Ownership			Foundational			
Single Source			Foundational			
Choosability			Control			
Standard			Sustainable	Sustainable		
Cost			Sustainable	Sustainable		
Availability			Security	Security		
Disclosure			Control	Foundational		
Validity				Security		

◇ render groups, while numbers define the instances of articles. For example, the group *SSI principles(15)* indicates nine articles concerning the principles of Self. The position of the diamond in the chart reflects the category of the journal in which the article was published. If most of the articles in the group pertain to technical journals, we aligned the diamond in the technical area of the chart. We used the *Scimago Journal & Country Rank*[2] and *Resurchify*[3] as a reference indicator to compare journals, and conferences, and help us categorize papers (when possible). The more articles published in technical conferences/journals, the more the diamond shifts toward the left hand. An in-depth chart analysis reveals two clusters. We interpret this as a tentative of SSI to combine definitions from players in different fields.

3   A Model to Assess Self-Sovereign Identity Solutions

Our model results from a two-step process as follows.

a. Existing definitions of SSI. From the systematic review study, we synthesized relevant works on SSI principles. Table 1 summarizes the results. The first column reports the principles. The subsequent columns reflect the name of the taxonomy for each work. The analysis of all works shows that the Sovrin Foundation (second column) gathers principles into a three-way taxonomy with Control, Portability, and Security (2016). Andrieu provides a tech-free categorization (2016) [11], and Ferdous extends principles to cover blockchain-based e-identity systems (2019) [12]. Gilani adds Validity as a further security property (2020) [13]. Others considered essential principles of Self (2019) [14,15].

We used past works to produce our definition of SSI, probing for similarities of snippets in the taxonomies. We paired snippets with similar intent and

---

[2] Scimago Journal & Country Rank. https://www.scimagojr.com/.
[3] Resurchify. https://www.resurchify.com/.

assigned principles to the category with the highest number of instances. We also omitted principles with duplicate meanings, namely Availability and Disclosure, ultimately obtaining a transitory table of twelve principles within their resulting category. To converge to a unique taxonomy, we designed four groups based on the following *what* and *how* questions: What are the fundamental human rights? What properties guarantee those rights? How can security be implemented? How can an e-identity scale up?

b. Producing the model. Our model results from a definition of *Challenges* and *Dimensions*. A challenge questions principles to bind theoretical properties and real-world initiatives. Then, a challenge produces dimensions to encode parameters. The next part details groups, principles and dimensions as reported in Table 2 and Table 3. The definition of the groups results from the previously mentioned *what* and *how* questions.

Individuals' rights. The category encloses principles for human rights.

- Existence: Individuals can assert attributes to services as proof of their identity. This principle includes assigned attributes that typically denote relationships with other entities, such as usernames and passwords [17]. We explore the option to generate new credentials from existing attributes. Additionally, we incorporate multi-factor authentication to finalize entity authentication following ISO 29115 guidelines [18]. We encode legal credentials (e.g., x509 and QWAC) to validate 'qualified' attributes according to CAdES specifications.
- Persistence. Individuals can present the same attributes from multiple issuers. Persistence distinguishes between Qualified and Non-Qualified Trust Service Providers that are comprehensively assessed under eIDAS. We also list private and public bodies that issue credentials without legal weight. Attributes may also be self-issued.
- Protection. It refers to the ability of systems to avoid censorship, ensuring that the list of identity and service providers is fairly managed.

Trustworthiness. The group encodes crucial features facilitating digital trust. It considers who can access the list of IdPs and attributes, and what attributes is possible to negotiate with an SP.

- Access. Access questions whether users can access the list of identity providers from a local wallet and get information about their attributes.
- Control. It refers to the possibility of individuals to negotiate attributes to a service provider through a user interface. It foresees the possibility of decoupling Personal Identification Data (PID) from other attributes.
- Transparency. Policies, rules, protocols and algorithms to manage the ecosystem members must be transparent. This refers to the possibility of assessing policies, algorithms, and software used to add/remove ecosystem entities.

Secrecy. The category frames properties for the Secrecy of information.

- Consent. It is the permission individuals give to collect, use, and share their data [19]. We consider the consent banner that appears as a pop-up to request user policy acceptance. Dynamic consent involves a dashboard to manage consent preferences. We included post-consent methods to manage consent preferences constraining the information flow [20].
- Data minimization. Users should only share the essential information with the service provider [21]. We explore options for individuals to selectively disclose information and consider the transfer of one attribute only or the associated information, for example, being over 18 years old.

Sustainability. The group advocates for the large-scale adoption of SSI.

- Cost. A digital identity system must be profitable for individuals, public and private organizations [12].
- Interoperability. Users can attest attributes across private and public services [22,23].
- Portability. Attributes can be transported to other ecosystems (GDPR Art. 20(1)(2)), between public and private services.
- Standard. An e-identity system must use globally recognized standards. We consider the readiness of stakeholders to include future standards reducing entry-level barriers through community groups, the industry sector, and government agencies, etc.

4   Assessment of VIDchain

This section provides a test bench for our model. We evaluate VIDchain, a Validated ID product that exploits the SSI paradigm's potential while issuing 'qualified' e-identities [4]. It complements the SSI-related specifications through software components that include the *VIDcredentials* to manage the creation and revocation of credentials. The *VIDwallet* organizes credentials, identifiers, and cryptographic keys. The *VIDconnect* is a custom implementation of the Self-Issued OpenID and did-auth protocol used to authenticate users towards a relaying party. Finally, the eIDAS Bridge allows issuing eIDAS-compliant certificates as verifiable credentials with legal weight [24]. During the evaluation phase, each dimension obtains a full ● mark to indicate that VIDchain complies with the dimension or an empty ○ mark if not. We assign half mark ◐ whenever the dimension is partially covered. Table 2 summarizes the evaluation of Individuals' rights and Trustworthiness. Table 3 summarizes the evaluation for Secrecy and Sustainability.

- *Existence.* VIDconnect supports authentication through username and password, along with multi-factor authentication. VIDcredentials creates a generic W3C Verifiable Credential in JSON-LD serialization format with LD signature (no legal value). However, the issuer signs the entire message, and the subject/holder cannot extract a single attribute from the JSON format to produce a new credential (empty mark).

Table 2. The list of principles, challenges and dimensions.

Individuals' Rights (a)			
Principle	**Challenge**	**Dimension**	**Eval.**
Existence	- What attributes can attest to an e-identity?	- Assigned attributes/ID tokens (Username and Password)	●
		- Multi-Factor Authentication (e.g., One-Time Password)	●
		- Combine attributes for a new credential	○
		- Legal credentials (e.g., x509/QWAC)	●
		- Other credentials (e.g., JWT-based, AnonCreds, ntQWAC)	●
		- Know Your Customer (KYC)	●
Persistence	- Who can issue attributes?	- Qualified Trust Service Providers (QTSPs)	●
		- Trust service providers (Non-Qualified)	●
		- Other public bodies (e.g., government agencies, Univ.)	◑
		- Other private bodies (e.g., Microsoft, Financ. Inst.)	◑
		- Foundations & intergovernmental organizations (IGOs)	○
		- Non-Governmental Organizations (NGOs) and others	○
		- Self-issued	○
Protection	- Who maintains the list of IdPs and SPs?	- Private sector (e.g., banks, credit bureaus)	○
		- Consortium of organizations (e.g., Kantara)	○
		- Government agencies (e.g., national identity authority)	●
		- Supranational organization (e.g., EU Commission)	○
		- Foundations & intergovernmental organizations (IGOs)	○
		- Open community of contributors/NGOs	○
		- Nobody	●
Trustworthiness (b)			
Principle	**Challenge**	**Dimension**	**Eval.**
Access	- How users obtain information about their attributes? - Can users access the list of IdPs?	- Local agent (wallet)	●
		- Shared ledger of IdPs	◑
		- History of attributes	○
Control	- Do users negotiate the release of attributes to SPs?	- User negotiates attributes but PIDs	○
		- User negotiates PIDs	○
		- Users can choose the service provider	◑
Transparency	- Are policies, rules, protocols and algorithms to manage ecosystem members open and clearly stated?	- Guidelines only	●
		- Transparent rules and procedures	●
		- Open protocols	○
		- Transparent algorithms	○
		- Open code/sftw	○
		- Open APIs	●

- *Persistence.* VIDchain can issue attributes provided by Qualified Trust Service Providers and Trust Service Providers under the eIDAS [3]. However, it accepts credentials from only a few private companies and a restricted number of financial institutions. The KYC onboarding process allows users to get government-issued credentials only from ID cards and passports. Although VIDchain produces credentials from phone numbers and email addresses, self-issued credentials, are not allowed.

- *Protection.* The Self-Issued OpenID Provider does not maintain a list of identity providers. However, under the eIDAS regulation, Member States establish, maintain and publish the trusted list of TSPs and QTSPs (Article 22 (1) of the eIDAS regulation).

- *Access.* There is no central map of trust service providers in the SSI ecosystem. However, the Commission holds it for TSPs and QTSPs in eIDAS[4]. Finally, there is no tool to track the history of attributes and those shared with services.

---

[4] EU/EEA Trusted List Browser. https://www.eid.as/tsp-map/#/.

Table 3. The list of principles, challenges and dimensions.

Secrecy (a)			
Principle	Challenge	Dimension	Eval.
Consent	- Does consent result adequately expressed and managed?	- Consent banner	●
		- Dynamic consent	○
		- Post-consent	○
Minimization	- Does the service lawfully collect only the minimum amount of information? - Do users employ techniques to limit data sharing?	- Selective disclosure of attributes but PIDs	○
		- Selective disclosure of PIDs	○
		- Transfer a new subset of attributes	○
		- Transfer one attribute at a time	○
		- Transfer the associated information only	○

Sustainability (b)			
Principle	Challenge	Dimension	Eval.
Cost	- To what extent does the e-identity is profitable for stakeholders?	- Profitable for public services	◖
		- Profitable for private services	◖
		- Profitable for citizens	◖
Interoperability	- To what extent can IdPs attest attributes to SPs across different jurisdictions?	- Among public services	◖
		- Among private services	◖
		- Among others (NGOs, IGOs, Found., etc.)	○
Portability	- To what extent can users transport the list of attributes on different ecosystems?	- Between public authorities	○
		- Between private authorities	○
		- Between others (NGOs, IGOs, Found., etc.)	○
Standard	- Who can issue standards for e-identity systems?	- Working/Community groups (e.g., W3C)	●
		- Industry sector (e.g., Okta)	●
		- Public agencies (e.g., NIST)	●
		- Other (Univ., NGOs, IGOs, Found., etc.)	●

- *Control.* There is no user interface to negotiate the release of attributes, and the choice of service providers is constrained to those accepting verifiable credentials.
- *Transparency.* The Self-Sovereign Identity (SSI) and the eIDAS define rules and the legal basis for ecosystem entities. Validated ID supplies APIs for all the endpoints. However, it is impossible to investigate the endpoints' source code.
- *Consent.* VIDchain utilizes the consent banner to ask for consent preferences. However, it does not implement different mechanisms to handle consent differently, such as dynamic or post-consent solutions that constrain the information flow.
- *Minimization.* In JSON-LD data is transformed and then hashed and then signed by the issuer with its private key. The holder can only present the entire credential as a verifiable presentation or non. This denies selective disclosure of attributes.
- *Cost.* The limited number of public/private institutions accepting verifiable credentials slows down the spread of this technology. On the other hand, VIDchain "bridges" SSI with eIDAS, thus opening a large ecosystem of approved identity/service providers.
- *Interoperability.* Interoperability among institutions is limited to a handful of accredited bodies and those who shifted to the new SSI paradigm by accepting VCs.

– *Portability.* The VIDwallet does not hold features for the export/import of credentials.
– *Standard.* Validated ID relied on SSI specifications, standards and recommendations. These standards and specifications are provided by many stakeholders, from Working/community groups, governments (eIDAS Bridge), W3C, DIF, Hyperledger Indy, Sovrin, ISA2 program, etc.

5   Discussion of the Results

VIDchain excels in safeguarding individuals' rights by attesting attributes in various serialization formats. The product accepts attributes from public and private identity providers; some of those services undergo major assessments under the eIDAS framework, and their list is free for everybody to browse. Notably, the absence of a centralized list of identity providers in SSI is also positive even though under the eIDAS, Member States maintain control over this list. Using the wallet guarantees end users reasonable control over their e-identities, and eIDAS contributes to robust interoperability across Europe.

   Concerns exist regarding the product's adherence to secrecy standards for end users. While it incorporates recommendations from various stakeholders, these standards may not adequately prioritize user secrecy of information. Verifiable credentials are signed by the issuer before issuance, namely the holder cannot combine those attributes forming a new credential and can either share every claim or non. This prevents selective disclosure of information. To address this, the wallet should implement features for attribute negotiation, allowing users to choose specific attributes to share. VIDchain also falls short in leveraging the Self-Sovereign Identity (SSI) paradigm for consent preferences and lacks a user-friendly dashboard for managing such preferences. Additionally, the absence of features for importing/exporting credentials hinders seamless movement across platforms.

6   Limitations and Conclusions

The study proposes a formalization to assess system adherence to Self-Sovereign Identity (SSI), using the VIDchain product as a test case. The model is intentionally general to avoid overfitting to specific solutions. As a consequent limitation, many implementation-specific parts were left blank for practitioners to fill in and were not tested by our model. For example, we generalized privacy-related issues of wallet authentication. Rapid technological advancement presents challenges in accurately anticipating all possible future scenarios, and the model may not consider new emerging technologies or unexpected changes in the technological landscape. Additionally, regulatory frameworks vary widely among jurisdictions, and regulation changes can significantly impact the scalability and adaptability of electronic identity systems. Therefore, incorporating the different regulatory contexts within which electronic identity systems operate may pose a challenge. At the current stage, the model does not address legal and compliance aspects that play a role in the implementation of electronic identity solutions. Lastly, the model lacks a user-friendly result, opting against a single score to prevent misconceptions. In summary, predicting the future technological landscape is challenging, and the model may not encompass all relevant factors, including emerging technologies, regulatory changes, and human-related aspects. Despite these limitations, the focus is on transitioning from theoretical principles to practical evaluation, marking the initial step toward creating a framework for assessing e-identity systems. Thus, future goals involve enhancing the model's quality by considering additional perspectives, testing in various jurisdictions, and incorporating industry initiatives to refine dimensions.

Funding. This work was partially supported by the European research projects H2020 LeADS (GA 956562) and Horizon Europe DUCA (GA 101086308), and CNRS EU-CHECK.

References

1. Yuxia, F., et al.: Non-transferable blockchain-based identity authentication (2023)
2. Allen, C.: The path to self-sovereign identity (2016). http://www.lifewithalacrity.com/2016/04/the-path-to-self-soverereign-identity.html. Accessed 21 July 2023
3. Regulation (EU) No 910/2014 of the European Parliament and of the Council of 23 July 2014 on Electronic Identification and Trust Services for Electronic Transactions in the Internal Market and Repealing Directive 1999/93/EC. https://eur-lex.europa.eu/legal-content/EN/TXT/?uri=CELEX:32014R0910. Accessed 29 Nov 2023
4. Vidchain. Vidchain Documentation. https://docs.vidchain.net/docs/intro. Accessed 21 Nov 2023
5. Schmidt, K., Mühle, A., Grüner, A., Meinel, C.: Clear the fog: towards a taxonomy of self-sovereign identity ecosystem members. In: 2021 18th International Conference on Privacy, Security and Trust (PST), pp. 1–7. IEEE (2021)
6. Satybaldy, A., Nowostawski, M., Ellingsen, J.: Self-sovereign identity systems: evaluation framework. In: Privacy and Identity Management. Data for Better Living: AI and Privacy: 14th IFIP WG 9.2, 9.6/11.7, 11.6/SIG 9.2. 2 International Summer School, Windisch, 19–23 August 2019, Revised Selected Papers 14, pp. 447–461 (2020)
7. Schardong, F., Custódio, R.: Self-sovereign identity: a systematic review, mapping and taxonomy. Sensors **22**(15), 5641 (2022)
8. Cushman, R., Froomkin, A.M., Cava, A., Abril, P., Goodman, K.W.: Ethical, legal and social issues for personal health records and applications. J. Biomed. Inform. **43**(5), S51–S55 (2010)
9. Badzek, L., Henaghan, M., Turner, M., Monsen, R.: Ethical, legal, and social issues in the translation of genomics into health care. J. Nurs. Scholarsh. **45**(1), 15–24 (2013)
10. Siddiqi, S., Sharan, A.: Keyword and keyphrase extraction techniques: a literature review. Int. J. Comput. Appl. **109**(2), 18–23 (2015)
11. Andrieu, J.: A Technology-Free Definition of Self-Sovereign Identity. Technical report (2016)

12. Ferdous, M.S., Chowdhury, F., Alassafi, M.O.: In search of self-sovereign identity leveraging blockchain technology. IEEE Access **7**, 103059–103079 (2019)
13. Gilani, K., Bertin, E., Hatin, J., Crespi, N.: A survey on blockchain-based identity management and decentralized privacy for personal data. In: 2020 2nd Conference on Blockchain Research and Applications for Innovative Networks and Services (BRAINS), pp. 97–101. IEEE (2020)
14. Sheldrake, P.: Generative identity - beyond self-sovereignty (2019). https://blog. akasha.org/generative-identity-beyond-self-sovereignty/
15. BlockTechDiVer. Potential of blockchain technology for digital consumer participation. https://blocktechdiver.de/. Accessed 30 Aug 2022
16. Tobin, A., Reed, D.: The inevitable rise of self-sovereign identity. Sovrin Found. **29**(2016), 18 (2016)
17. Preukschat, A., Reed, D.: Self-sovereign identity. Manning Publications (2021)
18. ISO/IEC 29115:2013 - Information Technology - Security Techniques - Entity Authentication Assurance Framework (2013). https://www.iso.org/standard/ 45138.html. Accessed 21 Nov 2023
19. Carolan, E.: The continuing problems with online consent under the EU's emerging data protection principles. Comput. Law Secur. Rev. **32**(3), 462–473 (2016)
20. Nissenbaum, H. (2018) https://hbr.org/2018/09/stop-thinking-about-consent-it-isnt-possible-and-it-isnt-right. Accessed 21 July 2023
21. Kilian, J., Micali, S., Ostrovsky, R.: Minimum resource zero-knowledge proofs. In: Brassard, G. (eds.) Advances in Cryptology. CRYPTO 1989. LNCS, vol. 435, pp. 545–546. Springer, New York (1989). https://doi.org/10.1007/0-387-34805-0_47
22. Yildiz, H., Küpper, A., Thatmann, D., Göndör, S., Herbke, P.: A tutorial on the interoperability of self-sovereign identities. arXiv preprint arXiv:2208.04692 (2022)
23. Choudhari, S., Das, S.K., Parasher, S.: Interoperable blockchain solution for digital identity management. In: 2021 6th International Conference for Convergence in Technology (I2CT), pp. 1–6. IEEE (2021)
24. Schwalm, S., Alamillo-Domingo, I.: Self-sovereign-identity and Eidas: a contradiction? Challenges and chances of Eidas 2.0. Wirtschaftsinformatik **58**, 247–270 (2021)

S.O.S. - My Grandparents: Using the Concepts of IoT, AI and ML for the Detection of Falls in the Elderly

Cosmin Rus[✉] [ID], Monica Leba [ID], and Remus Sibisanu [ID]

Department of System Control and Computer Engineering, University of Petrosani, 332006 Petrosani, Romania
cosminrus@upet.ro

Abstract. The increase in life expectancy and aging of the population have highlighted the need for innovative solutions to support and protect older people, who become more vulnerable to household accidents, particularly falls. This scientific work focuses on the development of an intelligent monitoring and alert system using Arduino Nano 33 IoT and TinyML technology to differentiate between falls and ordinary bending among the elderly. The system aims to promptly intervene in case of an unexpected event, ensuring access to medical assistance. The study explores the implementation of the system, including the use of advanced sensors and TinyML technology for accurate fall detection. Specialized literature on falls in the elderly and the potential of force platform measurements to predict falls is also reviewed. Ethical considerations and data privacy are emphasized throughout the research. The results show that the implemented system can accurately differentiate between sitting and falling movements, enabling rapid responses in emergency situations, thus contributing to improved elderly care and reducing the risk of complications associated with accidental falls.

Keywords: Arduino Nano 33 IoT · TinyML · machine learning · elderly fall detection

1 Introduction

In recent decades, the increase in life expectancy and the aging of the population have become increasingly relevant aspects in our society. As they age, older people become more vulnerable to household accidents, such as falls, which can have serious consequences for their physical and mental health. It is essential that we tackle this problem and find innovative solutions to support and protect older people as efficiently and quickly as possible. This scientific work focuses on the development of an intelligent monitoring and alert system, intended to differentiate between a fall of an elderly person and a simple bend, with the aim of promptly intervening in case of an unexpected event and ensuring access to medical assistance in the shortest possible time. This system has the potential to make a significant contribution to improving the quality of life of the elderly and reducing the risk of complications associated with accidental falls. To achieve this

Á. Rocha et al. (Eds.): WorldCIST 2024, LNNS 985, pp. 164–173, 2024.
https://doi.org/10.1007/978-3-031-60215-3_16

precise differentiation between a person's fall and an ordinary bend, we will explore various technologies and approaches. We will use advanced sensors and data analysis techniques to accurately detect the characteristic movements of falls and distinguish them from simple bending gestures. Through this work, we aim to develop a robust and scalable system that can be easily integrated into homes or care centers for the elderly. This system will be designed to operate autonomously, with the ability to notify the appropriate persons or emergency services in the event of a detected incident.

2   The Study of Specialized Literature

Falls among the elderly are a pressing global public health issue, significantly impacting their well-being and quality of life. Research in geriatrics and public health has emphasized the complexity of this problem, urging the search for effective solutions [1]. Rubenstein's comprehensive study in 2006 [1] delved into the epidemiology, risk factors, and preventive measures for falls among older adults, revealing that 30% to 60% of them experience falls annually, with 10%–20% leading to severe consequences, including hospitalization and mortality.

Identifiable risk factors such as muscle weakness, balance issues, confusion, and specific medications were associated with falls [1]. Addressing these factors substantially reduced fall occurrences, and targeted programs evaluating individual risks and environmental hazards proved effective [1]. Peel's work in 2011 [2] further highlighted falls' adverse effects on health, mortality rates, and healthcare costs globally, proposing public health approaches to manage fall risks efficiently.

Piirtola and Era's 2006 study [3] examined the force platform's ability to predict falls in the elderly. Certain parameters derived from force platform data, like center of pressure displacement, emerged as significant indicators of future falls, underscoring the need for more research in this area [3]. Recent works by Bargiotas et al. (2023) [4] and Dormosh et al. (2022) [5] have contributed insights into early fall risk detection and the development of predictive models for older adults. Implementing alarm systems for the elderly could offer prompt assistance during emergencies, fostering better communication between caregivers and the elderly [6, 7]. However, ethical considerations regarding privacy, consent, and dignity must be meticulously upheld [6–8]. These collective findings emphasize the urgency for further interdisciplinary research to develop more effective preventive strategies, ensuring the well-being of the elderly and enhancing their quality of life.

3   Materials and Methods

The elderly population is often at risk of falls and other incidents that can have serious consequences on their health. These situations can create anxiety and restlessness among caregivers, who want to ensure the safety and adequate care of their loved ones. Therefore, the development of an alarm and monitoring system becomes essential to meet the needs of these vulnerable people and to bring greater peace and security to both them and their relatives. In this context, the design and implementation of an innovative and efficient system becomes a necessity in providing adequate and prompt emergency assistance. Thus, we focus on the practical implementation of our innovative fall alarm system for the elderly using Arduino Nano 33 IoT and TinyML technology. This system is an efficient and accurate solution for monitoring and protecting the elderly, providing relatives and caregivers with a valuable tool to respond quickly in emergency situations. We will explore in detail the components required to implement the system, including the Arduino Nano 33 IoT and the IMU (inertial measurement unit).

The Arduino Nano 33 IoT is a compact and powerful development board that combines the capabilities of a microcontroller with Internet connectivity. It allows us to make an efficient interface between IMU sensors and TinyML technology. The IMU is a crucial component in detecting and differentiating intentional movements from accidental falls in the elderly. We will explore the functionality of the IMU and how it measures the acceleration and rotation of the device to provide relevant data. We will also introduce the concept of TinyML, an innovative approach to machine learning at the level of small and embedded devices. We will explore how we can use TinyML technology to train a model capable of recognizing intentional movements and accidental falls. In the implementation steps, we will detail the steps required to configure and program the Arduino Nano 33 IoT and IMU. We will also explore the process of training the TinyML model, including collecting and processing the training data, selecting the right algorithm, and tuning the parameters to get the best results. We will also discuss the integration of the system with a communication platform such as a mobile phone or a monitoring service. We will explore options for sending alerts to loved ones or emergency services to ensure a quick response in the event of a fall or emergency.

Finally, we will focus on evaluating the performance of the implemented system, analyzing its accuracy and efficiency in differentiating intentional movements from accidental falls. We will identify possible challenges and limitations of the system and discuss possible improvements and optimizations in future implementation. Overall, this implementation chapter is a crucial step in the development of our fall alarm system for the elderly. By using advanced technology and ethical principles, we aim to provide an effective and safe tool that contributes to the improvement of the quality of life of the elderly and the peace of mind of their relatives.

Designing an elderly fall alarm system using the Arduino Nano 33 IoT board and TinyML technology is an innovative and effective approach to ensuring safety and rapid response in emergency situations. This combination of hardware devices and tiny machine learning technology (TinyML) allows the development of a compact, intelligent system adapted to the specific needs of the elderly.

The Arduino Nano 33 IoT is a compact and powerful development board that integrates an ARM Cortex-M0 microcontroller and Wi-Fi and Bluetooth Low Energy (BLE) modules. TinyML technology is all about implementing machine learning models on resource-constrained devices like the Arduino Nano 33 IoT. By using these patterns, the alarm system can detect and recognize movements associated with a fall and trigger appropriate alerts.

4   Implementation, Results and Discussions

The implementation of the "S.O.S. - My Grandparents" system using the Arduino Nano 33 IoT board and the Edge Impulse platform is an innovative solution for the detection and clear differentiation between sitting and falling among the elderly. This system aims to ensure safety and quick assistance in the event of a fall, reducing the risk of accidents and providing a prompt response in emergency situations. By integrating IoT technology and Edge Impulse's machine learning capabilities, the "S.O.S. - My Grandparents" system becomes a smart and efficient solution for home healthcare. The clear differentiation between a sitting and a falling movement is a crucial aspect for the success of the whole "S.O.S. - My Grandparents" system. This precise distinction is essential to ensure the proper and effective functioning of the fall warning system, contributing to the provision of appropriate and prompt assistance to the elderly. The importance of the precise differentiation lies in the fact that the movement of sitting and falling have different meanings and implications for the user of the system. While a sitting motion represents a voluntary and normal action, a fall indicates an emergency situation and requires immediate intervention. The clear differentiation between the two types of movements allows the system to provide an appropriate and adapted response to the respective situation. By correctly identifying and labeling movements in the dataset used to train the machine learning model, the system's ability to accurately distinguish between sitting and falling is ensured. This step is essential to create an accurate and reliable model that can correctly detect and report critical events, thus ensuring safety and adequate support for users.

The first step in implementing the system is to collect data relevant to differentiating between sitting and falling. Sensors such as the accelerometer, gyroscope and magnetometer was used to record body movements and position. In Fig. 1 is represented in a graphic form the recording of specific data in the case of a voluntary sitting of an elderly person, while in Fig. 2 shows the data for the moment when a fall of this person is reported.

Fig. 1. Data obtained for the case when a person sits on a surface

Fig. 2. Data obtained for the case of a person falling on a surface

The retrieved data are then processed and prepared for the next stage of system implementation. In Fig. 3 shows the processing of the data taken for the case when a person sits on a certain surface and in Fig. 4 represents the data processing for the case that person falls on a certain surface.

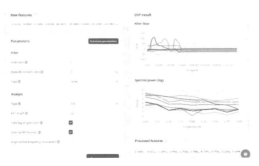

Fig. 3. Data processing for the sitting case

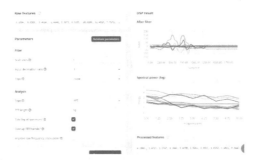

Fig. 4. Data processing for the falling case

After collecting and processing the data, it must be properly labeled to identify the moments of sitting and falling in the data set. Correctly labeling the data is essential for training the machine learning model (Figs. 5 and 6).

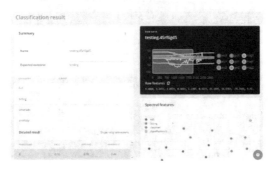

Fig. 5. Recording the moment of a sitting in the data set

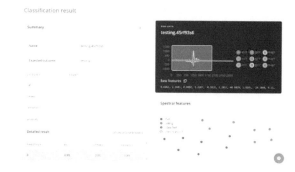

Fig. 6. Recording the moment of a fall in the data set

Using the Edge Impulse platform, a machine learning model was trained to differentiate between sitting and falling motions. Data collected from 3 subjects were used, 3 sets for each. Algorithms such as logistical regression and K-mean anomaly detection was used to achieve this differentiation (Fig. 7 and Fig. 8).

Fig. 7. Training the machine learning model

Fig. 8. Completing the training stage of the machine learning model

After training the model, it was validated using a separate test data set. The performance and accuracy of the model will be evaluated and optimizations was made to improve the results (in our case, 100% accuracy is observed regarding the clear differentiation between a fall and a settlement, Fig. 9).

Fig. 9. Machine learning model validation

There was used supervised anomaly detection, based on a labeled dataset containing both normal, that is sitting, and anomalous, that is falling, samples to construct the predictive model to classify the movements (Fig. 10).

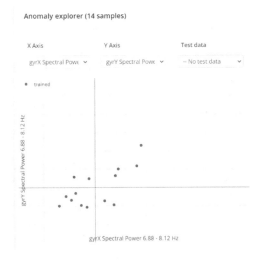

Fig. 10. Anomaly explorer

After the training and optimization of the model is completed, it was uploaded and implemented on the Arduino Nano 33 IoT board. The board takes the data from the sensors, process it using the model and issue an alert if a fall is detected.

The integration of a warning system into the structure of the system made with Arduino and TinyML adds significant value to the overall functionality, providing an immediate and effective notification in case of detection of a critical event, such as a fall of the elderly person. The purpose of this integration is to ensure a prompt reaction and alert the relevant people, such as family members, assistants or emergency services, to intervene appropriately and provide the necessary assistance in the shortest possible time. The warning system can be implemented through various devices and mechanisms suitable for fast and effective communication in emergency situations. Within this context, several options can be considered to be integrated into the system:

- SMS message or push notification system: Via an internet connection or a GSM module, the system can be configured to automatically send SMS messages or push notifications to the mobile phones of family members or responsible assistants. These notifications may contain relevant information about the detected event, such as date, time and location.
- Automated voice calls: Integrating an automated calling system can also be beneficial, allowing the system to directly call the designated phone numbers of responsible individuals. By means of pre-recorded voice messages, the system can clearly and quickly convey the information about the fall and the request for assistance.
- Integration with smart devices: If the warning system is connected to a smart network or digital assistance, notifications can also be delivered via other smart devices such as smartwatches or voice assistants.
- Sirens or local alarms: In addition to external notifications, the system can be equipped with sirens or local alarms, which emit a sound signal in the home or room to attract attention and alert the elderly person that assistance is on the way.
- Integration with emergency services: In serious situations where quick assistance is essential, the system can be configured to automatically alert local emergency services, providing them with essential information about the fall and location.
- This integration should be done carefully and in compliance with ethical and data privacy principles [11].

It is recommended that the system be configured with clear and well-defined notification options to avoid sending unwanted information or false alarms [12, 13].

Even if the main purpose of this work is not to present a practical realization of a warning system in its hardware construction that is particularly simple by integrating a GSM module into the Arduino ecosystem, the principle diagram is presented in Fig. 11. The work focuses on deepening and elucidating new concepts, especially in the complex functionalities and applications of artificial intelligence (AI) and machine learning (ML) technologies.

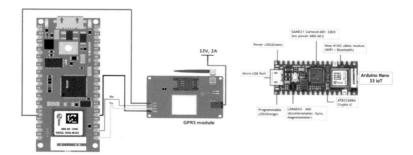

Fig. 11. Implementation scheme of the warning system using a GSM connection [14]

Instead of focusing primarily on hardware specifics, this paper approaches a method of extensive exploration of the revolutionary concepts and paradigms offered by AI and ML technologies, in a simple and easy-to-understand way. It strives to comprehensively present these cutting-edge concepts, elucidating their diverse applications and their role in addressing the complicated problem of fall detection and prevention in the elderly.

5 Conclusions

The "S.O.S. - My Grandparents" system is an innovative solution, based on the Internet of Things (IoT), for detecting falls and assisting the elderly. This scientific project addresses the need for effective and rapid solutions to support and protect older people, who become more vulnerable to domestic accidents, especially falls, as life expectancy increases. By using advanced technology, including the Arduino Nano 33 IoT board and TinyML technology, the system demonstrates accurate and reliable differentiation between sitting and falling motions. This accurate detection enables prompt intervention in the event of an unexpected event, triggering immediate assistance and access to medical support. By providing a rapid response to emergency situations, the system contributes to improving the care of the elderly and helps to reduce the risk of complications associated with accidental falls. The study of the specialized literature on falls among older people emphasizes the importance of the problem at the global level. Falls are a major public health concern, leading to serious injuries, hospitalizations, and even deaths among older adults. Rubenstein and Peel's research highlights the complexity of falls and identifies risk factors that can be addressed to reduce their frequency. Understanding the challenges and potential of force platform measurements to predict falls opens avenues for further research and prevention strategies.

The implementation of the "S.O.S. - My Grandparents" system using the Arduino Nano 33 IoT board and TinyML technology demonstrates a practical approach to solving the challenge of fall detection. The combination of compact and powerful hardware with machine learning capabilities allows the development of an intelligent and efficient solution adapted to the specific needs of the elderly. Integrating a warning system adds significant value to the overall functionality of the system by providing immediate alerts to relevant people or emergency services in the event of a fall. The implementation part demonstrates the steps involved in collecting and processing the data, training the machine learning model, and validating its accuracy. Achieving high accuracy in differentiating between falls and sittings confirms the effectiveness of the system. Ethical considerations and data protection are emphasized throughout the research, recognizing the importance of respecting the autonomy and dignity of older people. Proper education and communication with the elderly and their relatives is essential in building trust and promoting responsible use of the alarm system.

References

1. Rubenstein, L.Z.: Falls in older people: epidemiology, risk factors and strategies for prevention. Age Ageing **35**(suppl_2), ii37–ii41 (2006)

2. Peel, N.M.: Epidemiology of falls in older age. Can. J. Aging/La Rev. Can. Vieillissem. **30**(1), 7–19 (2011)
3. Piirtola, M., Era, P.: Force platform measurements as predictors of falls among older people–a review. Gerontology **52**(1), 1–16 (2006)
4. Bargiotas, I., et al.: Preventing falls: the use of machine learning for the prediction of future falls in individuals without history of fall. J. Neurol. **270**(2), 618–631 (2023)
5. Dormosh, N., Schut, M.C., Heymans, M.W., van der Velde, N., Abu-Hanna, A.: Development and internal validation of a risk prediction model for falls among older people using primary care electronic health records. J. Gerontol.: Ser. A **77**(7), 1438–1445 (2022)
6. Sundgren, S., Stolt, M., Suhonen, R.: Ethical issues related to the use of gerontechnology in older people care: a scoping review. Nurs. Ethics **27**(1), 88–103 (2020)
7. Pirzada, P., Wilde, A., Doherty, G.H., Harris-Birtill, D.: Ethics and acceptance of smart homes for older adults. Inform. Health Soc. Care **47**(1), 10–37 (2022)
8. Mittelstadt, B., Fairweather, B., Shaw, M., McBride, N.: Ethical issues of personal health monitoring: a literature review. In: ETHICOMP 2011 Conference Proceedings. Centre for Computing and Social Responsibility (2011)
9. Gitman, Y., Murphy, J.: PulseSensor on the arduino IoT cloud via the nano 33 IoT. In: Heartbeat Sensor Projects with PulseSensor: Prototyping Devices with Biofeedback, pp. 209–239. Apress, Berkeley (2023)
10. Abadade, Y., Temouden, A., Bamoumen, H., Benamar, N., Chtouki, Y., Hafid, A.S.: A comprehensive survey on TinyML. IEEE Access J. Neurol. **270**(2), 618–631 (2023)
11. Riurean, S., Rosca, S., Rus, C., Leba, M., Ionica, A.: Environmental monitoring systems in schools' proximity areas. In: Antipova, T., Rocha, Á. (eds.) MOSITS 2017. AISC, vol. 724, pp. 47–55. Springer, Cham (2018). https://doi.org/10.1007/978-3-319-74980-8_5
12. Rosca, S.D., Leba, M., Sibisanu, R.C., Panaite, A.F.: Brain controlled lego NXT mindstorms 2.0 platform. In: 2021 International Seminar on Intelligent Technology and Its Applications (ISITIA), pp. 325–330. IEEE (2021)
13. Panaite, A.F., Leba, M., Olar, M.L., Sibisanu, R.C., Pellegrini, L.: Human arm motion capture using gyroscopic sensors. In: MATEC Web of Conferences, vol. 343, p. 08007. EDP Sciences (2021)

Ethics and Security in the Era of Big Data: Innovative Challenges and Educational Strategies

Paola Palomino-Flores[1]([⊠]) [iD], Ricardo Cristi-Lopez[2] [iD], Edison Medina La Plata[3] [iD],
and David Paul[4] [iD]

[1] School of Communication of Universidad Peruana de Ciencias Aplicadas, Lima 15023, Peru
`paola.palomino@upc.edu.pe`
[2] School of Education of Universidad Andrés Bello, 8370134 Santiago, Chile
[3] Business School of Universidad Peruana de Ciencias Aplicadas, Lima 15023, Peru
[4] Business School University of Pennsylvania, Philadelphia, PA 003378, USA

Abstract. This study explores the integration of ethical and security principles within the Information Technology (IT) curriculum, focusing on Big Data. It examines key ethical issues like data privacy and management, alongside establishing secure handling guidelines for sensitive information. The research also identifies best practices for educational data protection and potential cybersecurity threats in academic settings. Using a mixed-methods approach combining qualitative and quantitative research, the study analyzes survey responses from 500 Systems Engineering students, offering insights into their perceptions of ethical and security training. The findings highlight the need for curriculum innovation that integrates ethical and security education, advocating for case studies and scenario-based learning to equip students with practical data management skills. Furthermore, the study emphasizes the urgency of IT education reform, stressing the inclusion of comprehensive ethical and security training. It also underlines the importance of academia-tech industry collaboration to align educational standards with industry demands, preparing graduates to navigate the ethical and security challenges in the digital world.

Keywords: Big Data · Ethics · Curriculum development

1 Introduction

The intersection of ethics and security in the realm of Big Data has become a pivotal concern in the Information Technology (IT) sector [1]. As Big Data continues to revolutionize information processing and analysis, it presents a multitude of ethical and security challenges for IT professionals. This complexity necessitates a reevaluation of ethical frameworks and security protocols to ensure they are robust enough to handle modern data environments [2, 3]. In this digital age, where data increasingly becomes a cornerstone of decision-making, the ethical and security challenges in Big Data management are becoming more complex and multifaceted [4–6]. Ethical theories such as consequentialism, deontology, and human nature theories play a significant role in shaping our understanding and response to these challenges [7]. These theories provide a critical lens for analyzing and addressing the ethical dilemmas and security concerns posed by Big Data. Big Data refers to the massive and diverse sets of data generated at an unprecedented scale, speed, and variety, typically measured in terabytes or more. It necessitates advanced technologies and analytical methods for capturing, storing, managing, and analyzing it [8]. To ensure compliance with data-protection laws, companies should implement encryption, authentication, and anonymization measures [9]. This data encompasses various formats, including structured and unstructured data from sources like sensors, social media, and online transactions, often requiring real-time or near-real-time processing to derive valuable insights. This complexity underscores the need for a comprehensive approach to IT education that prepares students to manage data responsibly and ethically [10].

The influence of Big Data on a global scale is profound, offering significant advantages to organizations that can effectively deploy it, though its full potential remains largely untapped [11]. IT professionals face the task of overcoming ethical and legal challenges, such as issues of data ownership, privacy, and security, to ensure compliance with regulations and to prevent valuable data from becoming a liability. For companies to fully benefit from big data, effective management of change across five key areas is essential: Leadership, Talent Management, Technology, Decision Making, and Corporate Culture [12]. Historically, businesses have been able to collect whatever data they wanted for any reason, with very little oversight. However, this scenario is beginning to shift as regulations are established to tighten the manner in which companies collect, store, and use data [13]. This highlights the importance of this study, aimed at enhancing IT professionals' ethical judgment to navigate these challenges. IT professionals are entering an era of increased organizational concern over data use. The advantages of Big Data are significant, with industries like banking, retail, telecommunications, government, science, and sports making considerable progress through its smart use. Yet, the shift from an open data environment to one with stringent data protection policies marks a critical transition. Consequently, this study recognizes the need to incorporate these ethical frameworks into IT curricula, aiming to equip future IT professionals with both theoretical and practical tools to navigate the ethical and security landscapes of the Big Data era. This approach is essential for developing a technically proficient, ethically informed, and security-conscious IT workforce capable of addressing the evolving challenges of the Big Data era. This study aims to address two primary objectives: firstly, to assess how current IT education programs integrate ethical theories and security practices in relation to Big Data, and secondly, to identify and propose effective strategies for enhancing the ethical and security preparedness of future IT professionals. In pursuit of these objectives, the study draws upon various ethical theories, including consequentialism, deontology, and human nature theories, as fundamental frameworks for ethical reasoning and practices in diverse contexts. The central research question guiding this study is: How can IT education programs be effectively structured to equip future IT professionals with the necessary ethical and security competencies to responsibly manage Big Data?

2 Background

2.1 The Ethical Landscape

The traditional definition of ethics, as elucidated by contemporary philosophers such as Robert C. Solomon and Johnson, positions ethics as a comprehensive system of theories that establish guidelines for moral decision-making [14]. These theories, deeply embedded in the concepts of value, virtue, and right action, derive from the ancient Greek word 'ethos', signifying character [15]. This etymological root highlights the intrinsic role of ethics in molding an individual's actions within the societal spectrum of good to bad, and right to wrong. Solomon and Johnson highlight the critical role of ethical considerations in shaping character and guiding decisions in IT and Big Data. They emphasize the need for a solid ethical framework to navigate the significant implications of rapid advancements in these fields [16].

In the realm of ethics, distinct perspectives offer varying guidelines for moral behavior. Ethical Egoism, particularly in its direct form, advocates for actions that primarily serve an individual's or an organization's self-interest, to the extent that profit maximization and potentially harmful actions, like forced labor or discrimination, could be deemed morally justified if they benefit the agent [17]. Throughout the centuries, various ethical theories have been developed, each offering distinct perspectives on judging human actions and guiding moral decisions. Consequentialism, for example, evaluates actions based on their outcomes, where a desirable result is indicative of a good action [18, 19]. Deontology presents a different approach, focusing on the will and duty behind actions rather than their consequences. In this theory, an action is deemed good or bad depending on the inherent will and reason behind it, a perspective that is especially significant in IT and Big Data [20]. The theory asserts that the intention and duty behind actions are critical in ethical judgments. Incorporating these ethical theories into the realms of IT and Big Data equips professionals with the necessary philosophical grounding to navigate the complex ethical landscapes they encounter. Understanding and applying these theories can guide IT professionals in making informed, responsible decisions, thus ensuring that their work aligns with ethical standards and principles [21].

2.2 Human Nature Theory in IT Education

The Human Nature Theory in IT education posits that people inherently have the faculties necessary for ethical behavior and decision-making. This approach asserts that, through targeted guidance and education, IT professionals can enhance their natural ethical inclinations. This becomes increasingly vital in the IT field, where rapid technological advancements and complex ethical challenges, including data privacy, cybersecurity, and the ethical deployment of artificial intelligence, are prevalent. By recognizing and cultivating these innate ethical strengths, education serves a crucial function in developing professionals who are not only adept in technology but also deeply ethical [23].

2.3 Developing Capabilities for Ethical IT Practice

The focus on developing capabilities for ethical IT practice involves a curriculum that goes beyond technical skills and delves into the moral and ethical aspects of technology use [24]. This approach is based on the idea that ethical behavior in IT is not just about adhering to external codes of conduct, but also reflects the internal ethical compass of professionals [25]. IT education should therefore focus on cultivating a deep understanding of ethical principles, promoting critical thinking, and encouraging students to consider the broader impact of their technological decisions [25]. Integrating case studies, ethical dilemmas, and discussions on real-world scenarios can help students apply theoretical ethical principles to practical situations, thereby strengthening their ability to make sound ethical judgments in their professional lives.

Understanding Human Nature Theory in IT focuses on each individual's capacity for ethical reasoning and action. It involves creating an educational environment that promotes self-reflection and moral reasoning, emphasizing the ethical implications of actions in IT [26]. Educators are crucial in helping students critically assess the ethical dimensions of technology and its impacts on society, culture, and individuals. This approach equips them to make decisions that are both technically sound and ethically informed [27]. The aim is to develop IT professionals with both technical prowess and ethical integrity, ensuring future leaders make beneficial decisions for business, technology, and society [28].

2.4   Application of Ethical Theories to Big Data Challenges

Integrating ethical theories into the IT curriculum is essential for addressing the unique ethical challenges of Big Data [29]. According to Solomon's definition in "Morality and the Good Life," ethics is grounded in values, virtues, and right actions, forming a foundational framework for ethical behavior in IT [14]. Ethical theories like consequentialism, deontology, and human nature theories provide valuable insights for ethical decision-making in Big Data security [30]. Consequentialism, which assesses actions based on their outcomes, guides IT professionals in evaluating the impacts of decisions on data privacy and security [10]. Forms of consequentialism, such as egoism, utilitarianism, and altruism, offer perspectives on prioritizing individual versus collective interests in Big Data. Deontology, focusing on duty and the motivation behind actions, provides a principles-based approach to decision-making, essential for handling sensitive data [31]. The human nature theory, viewing humans as inherently ethical, encourages IT professionals to foster responsible action in Big Data [22]. IT education must incorporate these ethical frameworks to develop students' comprehensive understanding of the moral implications of their actions in Big Data [32]. This requires a curriculum that teaches technical skills and ethical capabilities. Using Big Data case studies, the curriculum aims to develop professionals skilled in ethical decision-making, aligning with individual rights and societal norms [27, 28].

2.5   Curricular Innovations for Ethical Big Data Use

Integrating ethical aspects of Big Data into IT education is crucial. This includes teaching ethical theories like consequentialism, which assesses actions based on outcomes, and deontological theories that focus on intentions and duties [10]. Such integration helps students understand Big Data's ethical complexities and prepares them for responsible decision-making. Collaboration between educational institutions and the tech industry is essential to enhance ethical training [33]. These partnerships provide real-world examples and practical applications of ethical theories. Students gain hands-on experience, preparing them for future ethical challenges in areas like privacy and data use. It's important to align educational practices with industry standards and legal requirements in data handling. Teaching compliance, data governance, and the implications of non-adherence fosters a culture of ethical responsibility and legal compliance in the future IT workforce [10, 33].

3 Methodology

The study employs a mixed-methods approach to assess the integration of ethical theories and security practices in IT education [34]. Initially, it conducts a qualitative analysis of the Systems Engineering curricula at three universities in Peru. This phase critically reviews course content, teaching materials, and the overall structure of the programs, aiming to identify the strengths and weaknesses in addressing ethical and security issues related to Big Data. Following this, the study progresses to a quantitative phase, surveying 500 students from three private universities [35]. These students are selected based on criteria such as their year of study, diversity of interests, and involvement in Big Data projects. The surveys are designed to measure students' perceptions of the effectiveness and relevance of their ethical and security training in IT.

Data for this study were collected via an internet-based survey, with all participants providing digital informed consent [36]. The consent form presented the study's aims, research methods, data collection and storage processes, access to data, and highlighted the voluntary nature of participation. Analysis of the gathered data was conducted using SPSS statistical software. By combining qualitative curriculum analysis with quantitative survey data, the study adheres to ethical research standards, acknowledging potential biases in participant selection and the scope of the institutions involved. The anticipated outcome is to inform the development of more effective educational strategies and curricula that respond to the evolving ethical and security challenges in the IT industry.

4 Results

4.1 Curriculum

The comprehensive analysis of Systems Engineering curricula at three private universities revealed varying degrees of integration of ethical theories and security practices. While some programs demonstrated a strong emphasis on ethical aspects, particularly in relation to data privacy and cybersecurity, others showed significant gaps, particularly in the practical application of these theories in the context of Big Data. The review identified a need for more structured and comprehensive modules that specifically address the ethical and security challenges of Big Data.

4.2 Survey Results

From the five hundred surveys collected from students enrolled in Systems Engineering programs, several key trends emerged. The majority of students acknowledged the importance of ethical and security training in their curriculum, yet many indicated a lack of sufficient coverage of these topics. Particularly, students expressed a desire for more hands-on experiences and case studies that link ethical theories to real-world Big Data scenarios. The survey results also indicated a general awareness among students of the ethical implications of Big Data, but a gap in their ability to apply this understanding to practical situations.

4.3 Integration of Ethical Theories and Security Practices

The data suggested that while ethical theories are mentioned in the data suggested that while ethical theories are mentioned in the curriculum, their integration with practical security measures in Big Data is not always evident. This indicates a disconnect between theoretical ethical frameworks and their application in the rapidly evolving field of Big Data.

4.4 Preparedness for Future Challenges

The results show that students feel moderately prepared to face future ethical challenges in Big Data. However, there is a clear need for more focused training that equips them with both the theoretical knowledge and practical skills to navigate complex ethical and security landscapes (Fig. 1).

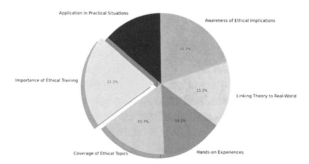

Fig. 1. Survey results from Systems Engineering students, showing their perceptions on various aspects of ethical and security training in their curriculum

The pie chart visualizes survey results from Systems Engineering students about their perceptions of ethical and security training in Big Data. Each chart segment, varying in size based on a 1 to 5 rating scale, represents a different aspect. The slightly exploded segment for the Importance of Ethical Training shows its high value among students. The yellowgreen segment for Coverage of Ethical Topics indicates a moderate perception of curriculum coverage. Hands-on Experiences, in lightcoral, reveals a smaller proportion, suggesting a need for more practical curriculum elements. The lightskyblue segment on Linking Theory to Real-World scenarios displays moderate satisfaction. Awareness of Ethical Implications is represented by a larger orange segment, indicating good student awareness. The purple segment for Application in Practical Situations points to an area needing improvement. Overall, this chart shows strengths and areas for improvement in students' Big Data ethics and security education, highlighting Ethical Training and Awareness as strengths, and identifying Hands-on Experiences and Practical Application as areas needing development.

5 Discussion

In this critical discussion of the study, we delve into the integration of ethical frameworks and security practices in IT education, focusing on the challenges presented by Big Data. Solomon and Johnson's perspective on ethics highlights the significance of moral decision-making in the rapidly advancing field of IT and Big Data, aligning with the study's emphasis on ethical theories such as consequentialism, deontology, and human nature theories [2, 3, 6, 7]. In the study, the seamless integration of consequentialism [7], deontology [17], and human nature theory [19] in IT education forms a robust ethical framework for Big Data. Consequentialism guides students to assess the outcomes of their decisions in data-related scenarios, while deontology reinforces the importance of ethical rules and duty, crucial for data privacy and security. Bridging these theories, human nature theory encourages the cultivation of innate ethical faculties, enabling intuitive ethical decision-making. This cohesive approach [31, 32] ensures that IT professionals are well-equipped with a comprehensive understanding of various ethical principles, essential for navigating the complex landscape of Big Data.

There are significant obstacles related to data ownership, privacy, and security that must be overcome for any company to fully leverage data. Ignoring these issues or addressing them incorrectly could turn data from a vital asset into a potential and dangerous burden. The curriculum analysis and student surveys reveal a gap in the practical application of these ethical theories within Big Data contexts [2, 4], indicating a pressing need for educational innovation that effectively bridges theoretical ethics with real-world application [1, 4]. The moderate level of student preparedness for future ethical challenges in Big Data underscores a demand for curricula that provide both theoretical knowledge and practical skills [1, 19]. While students recognize the importance of ethical training, the findings point to a substantial need for focused training that integrates ethical theory with practical skills. The study's methodology, combining qualitative and quantitative approaches, provides a comprehensive understanding of the current state of IT education in relation to Big Data ethics and security, highlighting the urgent need for curriculum enhancements that address hands-on experiences and the practical application of ethical theories [31, 32].Conclusively, the study critically assesses the integration of ethical theories and security practices in IT education, shedding light on the significant gap in their practical application. The insights gained from the curriculum analysis and student surveys offer valuable directions for future educational strategies. There is a clear need for curriculum innovations that not only impart ethical knowledge but also facilitate practical experiences and real-world applications. Such an approach ensures the development of an IT workforce that is technically adept, ethically informed, and prepared to meet the evolving challenges of the Big Data era [7, 30].

6 Conclusions

The study "Ethics and Security in the Era of Big Data" underscores a critical gap in IT education, particularly in the integration of ethical frameworks into practical applications within Big Data contexts. Although Systems Engineering students are aware of ethical implications in Big Data, there is a notable discrepancy between their theoretical understanding and practical application skills, especially concerning security and privacy. This gap reveals a fundamental flaw in current educational models, where theoretical concepts are not sufficiently complemented with practical, hands-on experiences. Additionally, the study observes that current curricula underutilize human nature theories, which suggest that individuals inherently possess faculties for ethical action. This underutilization represents a missed opportunity in nurturing students' innate ethical potentials, essential for developing responsible IT professional's adept at navigating Big Data's ethical complexities. The study highlights limitations such as selection biases and a narrow focus, recommending a more holistic IT curriculum that incorporates simulations and ethical discussions to link theory with practice. It suggests stronger academia-industry partnerships for curriculum relevance and emphasizes enhancing students' ethical skills, as per human nature theories, to marry technical skill with ethical awareness. The study also advocates for continual curriculum updates to address evolving ethical challenges in Big Data, ensuring relevance for future IT professionals.

Acknowledgements. The authors would like to thank the participants who participated in this research and the partially supported by the Universidad Peruana de Ciencias Aplicadas (UPC).

References

1. Lyon, D.: Surveillance, snowden, and big data: capacities, consequences, critique. Big Data Soc. **1**(2), 205395171454186 (2014). https://doi.org/10.1177/2053951714541861
2. Aldboush, H.H.H., Ferdous, M.: Building trust in fintech: an analysis of ethical and privacy considerations in the intersection of big data, AI, and customer trust. Int. J. Financ. Stud. **11**(3), 90 (2023). https://doi.org/10.3390/ijfs11030090
3. Howe, E.G., III, Elenberg, F.: Ethical challenges posed by Big Data. Innov. Clin. Neurosci. **17**(10–12), 24 (2020)
4. Jain, P., Gyanchandani, M., Khare, N.: Big data privacy: a technological perspective and review. J. Big Data **3**, 25 (2016). https://doi.org/10.1186/s40537-016-0059-y
5. Méndez, E., Sánchez-Núñez, P.: Navigating the future and overcoming challenges to unlock Open Science. In: Ethics and Responsible Research and Innovation in Practice, pp. 203–223. Springer, Cham (2023). https://doi.org/10.1007/978-3-031-33177-0_13
6. Kempeneer, S.: A big data state of mind: Epistemological challenges to accountability and transparency in data-driven regulation. Gov. Inf. Q. **38**(3), 101578 (2021). https://doi.org/10.1016/j.giq.2021.101578
7. Kizza, J.M.: Ethical and social issues in the information age. Springer, Cham (2010). https://doi.org/10.1007/978-3-319-70712-9

8. Oussous, A., Benjelloun, F.-Z., Ait Lahcen, A., Belfkih, S.: Big Data technologies: a survey. J. King Saud Univ. – Comput. Inf. Sci. **30**(4), 431–448 (2018). https://doi.org/10.1016/j.jks uci.2017.06.001

9. Beg, S., Khan, S.U.R., Anjum, A.: Data usage-based privacy and security issues in mobile app recommendation (MAR): a systematic literature review. Libr. Hi Tech. **40**, 725–749 (2022)

10. Sarker, I.H.: Data science and analytics: an overview from data-driven smart computing, decision-making and applications perspective. SN Comput. Sci. **2**, 377 (2021). https://doi. org/10.1007/s42979-021-00765-8

11. Medina, E.: Big Data, los datos como generadores de valor (2023)

12. Harvard Business Review. Big data: The management revolution (2012)

13. Marr, B.: Data strategy. Teell Editorial (2018)

14. Buchanan, A.: [Review of Ethics and Excellence: Cooperation and Integrity in Business, by R. C. Solomon]. J. Bus. Ethics **13**(2), 94–154 (1994). http://www.jstor.org/stable/25072509

15. Carr, D., Arthur, J., Kristjánsson, K. (eds.): Varieties of Virtue Ethics. Palgrave Macmillan UK (2017)

16. Van Stekelenburg, L.H., Smerecnik, C., Sanderse, W., De Ruyter, D.J.: What do you mean by ethical compass? Bachelor students' ideas about being a moral professional. Empiric. Res. Vocat. Educ. Train. **12**(1) (2020). https://doi.org/10.1186/s40461-020-00097-6

17. James, H.S., Rassekh, F.: Smith, Friedman, and self-interest in ethical society. Bus. Ethics Q. **10**(3), 659–674 (2000). https://doi.org/10.2307/3857897

18. Darwell, S.: Deontology. Blackwell Publishing, Oxford (2002)

19. Guinebert, S.: How do moral theories stand to each other? ZEMO **3**, 279–299 (2020). https:// doi.org/10.1007/s42048-020-00077-1

20. Aboodi, R., Borer, A., Enoch, D.: Deontology, individualism, and uncertainty: a reply to Jackson and Smith. J. Philos. **105**(3), 259–272 (2008)

21. DeTienne, K.B., Ellertson, C.F., Ingerson, M.-C., Dudley, W.R.: Moral development in business ethics: an examination and critique. J. Bus. Ethics **170**(3), 429–448 (2021). https://doi. org/10.1007/s10551-019-04351-0

22. Fullagar, S.: Nature, performing. In: International Encyclopedia of Human Geography, pp. 305–308. Elsevier (2009)

23. Hawkins, D.: Human nature and the scope of education. Teachers College Record (1970) **73**(5), 287–326 (1972). https://doi.org/10.1177/016146817207300512

24. Meyer, M.W., Norman, D.: Changing design education for the 21st century. She Ji J. Design Econ. Innov. **6**(1), 13–49 (2020). https://doi.org/10.1016/j.sheji.2019.12.002

25. Giorgini, V., et al.: Researcher perceptions of ethical guidelines and codes of conduct. Account. Res. **22**(3), 123–138 (2015). https://doi.org/10.1080/08989621.2014.955607

26. Kaushik, V., Walsh, C.A.: Pragmatism as a research paradigm and its implications for social work research. Soc. Sci. (Basel, Switzerland) **8**(9), 255 (2019). https://doi.org/10.3390/soc sci8090255

27. Falloon, G.: From digital literacy to digital competence: the teacher digital competency (TDC) framework. Educ. Tech. Res. Dev. **68**, 2449–2472 (2020). https://doi.org/10.1007/s11423-020-09767-4

28. Ng, D.T.K., Leung, J.K.L., Su, J., et al.: Teachers' AI digital competencies and twenty-first century skills in the post-pandemic world. Educ. Tech. Res. Dev. **71**, 137–161 (2023). https:// doi.org/10.1007/s11423-023-10203-6

29. Atenas, J., Havemann, L., Timmermann, C.: Reframing data ethics in research methods education: a pathway to critical data literacy. Int. J. Educ. Technol. High. Educ. **20**, 11 (2023). https://doi.org/10.1186/s41239-023-00380-y

30. Bednar, K., Spiekermann, S.: The power of ethics: uncovering technology risks and positive value potentials in IT innovation planning. Bus. Inf. Syst. Eng. (2023). https://doi.org/10. 1007/s12599-023-00837-4

31. Frecknall-Hughes, J., Moizer, P., Doyle, E., Summers, B.: An examination of ethical influences on the work of tax practitioners. J. Bus. Ethics **146**(4), 729–745 (2017). http://www.jstor.org/stable/45022347
32. Guan, X., Feng, X., Islam, A.A.: The dilemma and countermeasures of educational data ethics in the age of intelligence. Hum. Soc. Sci. Commun. **10**, 138 (2023). https://doi.org/10.1057/s41599-023-01633-x
33. Card, D., Smith, N.A.: On consequentialism and fairness. Front. Artif. Intell. **3** (2020). https://doi.org/10.3389/frai.2020.00034
34. Cheong, H., Lyons, A., Houghton, R., Majumdar, A.: Secondary qualitative research methodology using online data within the context of social sciences. Int. J. Qualit. Methods **22** (2023). https://doi.org/10.1177/16094069231180160
35. Polkinghorne, D.E.: Qualitative research. In: Handbook of Clinical Psychology Competencies, pp. 425–456. Springer, New York (2010)
36. Grady, C., Cummings, S.R., Rowbotham, M.C., McConnell, M.V., Ashley, E.A., Kang, G.: Informed consent. N. Engl. J. Med. **376**(9), 856–867 (2017)

Human-Computer Interaction

Farming and Automation. How Professional Visions Change with the Introduction of ICT in Greenhouse Cultivation

Silvia Torsi[1]([⊠]), Luca Incrocci[2], Stefano Chessa[1], Alexander Kocian[1],
Paolo Milazzo[1], Fatjon Cela[2], and Giulia Carmassi[2]

[1] Department of Computer Science, University of Pisa, Pisa, Italy
`silvia.torsi@unipi.it`
[2] Department of Agriculture, Food and Environment, University of Pisa, Pisa, Italy

Abstract. With this contribution, we would like to contextualize an IoT and Artificial Intelligence project in the current work practices of greenhouse growers. The project follows the user-centered design process in the field of Human-Computer Interaction. We carried out user studies by interviewing experts competent in greenhouse work. In particular, we asked ourselves how practices, perceptions, and cognitions change following the introduction of technologies in the greenhouse. The qualitative work on the interviews allowed us to isolate some constants that will subsequently bring us to improve the usability of the technologies to support growers, taking into account their particular professional skills.

Keywords: Human-Computer Interaction · User Centered Design · User Interviews · Greenhouse culture · Professional Visions · ICT4Agriculture

1 Introduction

How does the practice of greenhouse cultivation change with the introduction of technologies? In recent decades, technological advancements in agriculture have reached dizzying speeds. However, the introduction of Computer Science in the greenhouse requires a shift in the paradigms, in the cognitive processes of the farmers. We approached this issue with 11 interviews [4] which saw academics from the Faculty of Agriculture of the University of Pisa (Italy) in the role of experts [8].

2 Rationale

The history of agriculture has been intertwined with that of humankind for millennia. Agriculture is part of its cultural repertoire in many ways, and its progress in man's history have represented one of the possible historical-material conditions that gave birth to civilizations that have now disappeared and have left us art, culture, science, law, philosophy, and technique. For many millennia, technical advances in agriculture have meant more health, quality of life, well-being, and wealth. In this contribution we

Á. Rocha et al. (Eds.): WorldCIST 2024, LNNS 985, pp. 187–193, 2024.
https://doi.org/10.1007/978-3-031-60215-3_18

want to focus on the sub-sector of growing (irrigation in particular) in the ecosystem of greenhouses cultivations. Manual work in the greenhouse requires constant dedication, seven days a week, which, over the years, and with the experience, determines in the grower a set of professional visions [5] that allow him to diagnose possible alterations very easily in the crop, potential threats and to decide with appropriateness the correct practices [14] to be carried out. Experienced farmers (particularly mature adults) thus apply their perceptions and problem-solving strategies as a daily activity. However, with the advent of sensors and artificial intelligence, these skills change and must be replaced by other knowledge and different practices. There is therefore a generational, cultural and practice gap that is occurring in the practices of plant growing. Within our project, we wanted to exploit this situation and analyze it in detail to better contextualize our desire to contribute to greenhouses management in terms of ICT and AI [12, 2].

3   Related Work

Being located at the intersection of ICT and Cognitive Science, we will briefly point out the main contributions in both fields. From the perspective of Computer Science, there has been a recent surge of interest in integrating agriculture with ICT research, given the vast potential this field holds [13, 18] has renamed this interest by paving the way for what is now called "ICT4Agriculture". One of the most prolific is the attention to agriculture in developing countries, HCI4D (human-computer interaction for development, that could particularly benefit from ICT [10, 19]. As for the introduction of technologies in agriculture, [15] for example highlights the professional visions described by [5] and cultivation practices as a cultural resource of agricultural communities that can be remapped on the light of ICT.

We start from the concept of artifact in Activity Theory (e.g. [3]), in which tools have a dual nature: the ones incorporated in the artifact and the ones corresponding to the mental structure of the farmer, built through settled practice and cultural transmission. We therefore carried out interviews with a set of agricultural academics in the role of informants to make the point of the situation. We started without preconceptions and asked general questions to probe cognition and practices in both manual and automated irrigation in the greenhouses.

From the perspective of cognition and practices, Reflexivity in action [7] is a significant component of professional practice. [7] describes it as the perception of patterns in a problem set that reveals the steps to carry on to solve the problem. The first and most comprehensive theoretical reference is the concept of professional visions in [5], who makes the point on the specific practices of each professional group for shaping events to be able to analyze them.

4   Background

There is a vast landscape in the field of automation for the activities in greenhouses. The most substantial is in the field of controlling the irrigation and fertilizations of the plants. Sensors are adopted for the most of the physical and chemical parameters of the overall ecosystem. Between direct and mediated processes of irrigation there is the widespread adoption of timers, representing the current standard to overcome with ICT. The automatic calculation of interventions over the plants from the sensors data provides a scientific approach that ensure the fitness of the greenhouse system and an important reduction of workload for the farmer.

5 Theoretical Framework

Professional practices consist of dividing the phenomenal world into specific cognitive objects of the profession and, in their manipulation, aimed at bringing out patterns that correspond to the mental structures typical of the profession. This manipulation involves either movement [20], cognition [7], or the partitioning of the perceptual field into figure and ground [20] to bring out structures and representations codified by the professional group and transform the space of the problem into something simple and evident that by itself suggests the solution procedure. All these activities represent the repertoire of a profession and are transmitted socially. Each profession represents a culture, which consists both in knowing how to see, in knowing how to act, in learning how to reason, and in knowing how to do. This ability to identify the significant elements within the space of a problem derives from the professional social context of belonging. The professional vision, therefore, consists of the encounter between the "out there" phenomenal world and the specific coding schemes of the profession, which transforms the multifaceted complexity of nature into the phenomenal categories that constitute the working environment of a profession. And through this encounter, nature is transformed into culture. The research question is: how is this landscape of skills modified by the introduction of ICT in the greenhouses?

6 The Interviews

We recruited for an interview 11 academics from the agricultural sciences department of our university (8 professors, 1 technician and 2 doctoral students). We asked them questions about the practices of silk growers whether they used automation methods to understand the differences. We attempted to delve deeper into the topic of the growers' ways of thinking by posing additional questions to better understand this aspect with and without automation. The interviews included the evaluation of two prototypes of web applications for the automatic management of irrigation as well as the monitoring of sensors (humidity, radiation, salinity, etc.). Additionally, questions were asked to elicit possible new ideas for soilless cultivation, and opinions were sought regarding the use of bacteria to combat parasites. The interviews took place in the of the Department of Agriculture. The interviews were transcribed, coded and the Grounded Theory analysis method [6] was applied to formulate abstractions and generalizations from the answers. The most interesting results on which a subsequent phase of design and implementation of technologies to support soilless cultivation practices can be based are presented below.

6.1 Results

All those interviewed experts gave the reduction of waste of water and fertilizers, the saving of time and cognitive resources as the first benefit of automation, and the majority named the risk of failures and the consequent cessation of automatic plant care as negative sides. More precise questions were asked about how growers' cognition changes with automation. The main outputs to these questions are described below.

From Heuristic to Scientific Reasoning. Introducing sensing and computation in irrigation processes first determines a change in the routines. The expert greenhouse grower is a person who, through experience, has refined his perception of the plant in such a way as to immediately understand if there is something wrong and move on to the remedy.

P11: "I think it is only a very small percentage that does not exceed 10% of people who, let's say, establish the amount of water to give and when to give it in a scientific way. Most farmers do it empirically by evaluating the plant visually or with simple methods such as touch to feel if the soil is too wet or dry."

P7: "Automation, on the other hand, requires monitoring of sensors through the parameters of a computer or smartphone. For example, they have to consider the plant both on the quantity of water supplied and on the regular growth of the plant. Atmospheric phenomena, namely the percentage of humidity in the air or the amount of irradiation, must be analyzed to verify the correct procedures. Factors such as the amount of drainage or the salinity of the soil come into play."

Together, all these parameters require a complex and varied evaluation throughout the day and the year. Therefore, the growers' reasoning shifts from heuristic and experience-based reasoning to a scientific and deductive attitude of many chemical and physical parameters to determine the right interventions on the plant.

Novel Perceptual Routines. Another parameter concerns the correct functioning of the automated system. The grower needs to integrate his diagnostic reasoning on the vegetation with a checking that the technological and mechanical components are effectively working.

P11: "Like all machines, a security system would be needed, because if the machine doesn't work, you have to be able to bypass the system manually. And above all [we would need] a system that allows you to warn yourself if it doesn't work. Because after you have become familiar with a system, which works on its own, you trust it to do so always and anyway, so you could relax your surveillance. And so if the system breaks for some reason, it stops because all machines can stop or break. You risk having major damage because it's not that you didn't have that vigilance."

Therefore, automation requires a shift in perception, in the direction of monitoring its functioning. For this reason, the perceptual processes of the farmers are restructured, encompassing the direct and indirect visual ad auditorial cues that the automation system provides.

The Cognitive Processes. The eye-hand-brain routines are also involved in this change. The grower must restructure his practice by embracing automation, with its different affordances, procedures to abandon, or activities to carry on in different ways. An interviewee explains this shift with the example of driving for the first time an electric car.

P5 "In my opinion, in the case of automation, however, the grower must have the experience of automation, because then if there is something wrong, he must be able to understand it. So, it's a different experience. It's a different experience, maybe I'll give you a comparison. So, let's say learned to drive a car. Ok now there is an electric car and without gearbox I happened to get it as a replacement. Everything seemed different to me, but then in the end I had to learn new things and maybe the car doesn't have the key in my hand. But it's not that I don't do anything anymore, and therefore I imagine a person who learns new mechanisms, including reasoning, but he does them, it's not that he stops doing them. That is, if I turn the key and insert the key and start the car, [with the electric car] instead I won't have to make that movement, maybe I just pass my hand in front of the door and the car opens. I imagine something like this."

Consequently, automation requires restructuring the professional skills, where there is a different way to work, and the need of modifying the cognitive processes in the direction of including the information coming from sensors and the functioning of the actuators in the overall practices and routines.

6.2 Discussion

It is possible to summarize the differences between manual and automated practices while considering the mediation of technologies. The farmers no longer operate directly on their ecosystem. Every perceptive, cognitive and coordination activity acts with and through ICT. This additional level acts as a mediator between the grower and his work environment. There is shift from experience and heuristic reasoning and acting toward a more systematic, objective, and scientific based practices. The improvements are substantial, but they need a change in most of the routines that not all the farmers would be willing to make. For this reason, there is the need to focus on usability and intuitiveness in all interfaces between the grower and the automated system while following the professional skills of the farmer as much as possible in order to minimize the alterations he needs to make to his practices. Hence, an important research work in cognitive anthropology and Human-Computer Interaction is essential to determine the adoption of technologies in the greenhouses.

7 Conclusions

The skills, practices, and knowledge of farmers are unique. Replacing farmers' heuristics with procedures based on science, computation, and engineering can substantially innovate agriculture practices. Anyway, the introduction of ICT in greenhouses cannot ignore this universe of farmers' skills. Cultural factors and consolidated practices could affect their acceptance and adoption. Analyzing the differences between the two approaches provides the basis for systematic research on the professional skills of farmers and concerns on how to incorporate this heritage into greenhouse innovation processes.

Acknowledgments. We would like to warmly thank Luca Botrini, Chiara Trimarchi, Rita Maggini, Paolo Vernieri, Alberto Pardossi, Alice Trivellini, Ferdinando_Malorgio, Martina Puccinelli for their time, their clarity and original contribution. This work is part of the following programs: Programma di sviluppo rurale 2014–2020. "Fondo europeo agricolo per lo sviluppo rurale: l'Europa investe nelle zone rurali." Programma di Sviluppo Rurale 2014–2020 – LIGURIA Misura M16.01 "Aiuti per la costituzione e l'operatività dei gruppi operativi del PEI" 2a fase– "settore

AGRICOLO" "Attuazione del progetto dei Gruppi operativi" (attuazione DGR n. 727/2019). This study was carried out in part within the Agritech National Research Center and received funding from the European Union Next-Generation EU (PIANO NAZIONALE DI RIPRESA E RESILIENZA (PNRR) – MISSIONE 4 COMPONENTE 2, INVESTIMENTO 1.4 – D.D. 1032 17/06/2022, CN00000022). This manuscript reflects only the authors' views and opinions, neither the European Union nor the European Commission can be considered responsible for them.

References

1. Abras, C., Maloney-Krichmar, D., Preece, J.: User-Centered Design. W. Encyclopedia of Human-Computer Interaction. Sage Publications, Thousand Oaks, In Bainbridge (2004)
2. Kocian, A., Carmassi, G., Cela, F., Chessa, S., Milazzo, P., Incrocci, L.: IoT based dynamic Bayesian prediction of crop evapotranspiration in soilless cultivations. Comput. Electr. Agricul. **205** (2023)
3. Bertelsen, O.W., Bødker, S.: Activity theory. In Carroll, J.M. (ed.). HCI Models, Theories, and Frameworks: Toward a Multidisciplinary Science, pp. 291–324. Elsevier (2003)
4. Bogner, A., Menz, W.: The theory-generating expert interview: epistemological interest, forms of knowledge, interaction. Interviewing experts. Research Methods Series, pp. 43–80. Palgrave Macmillan, London (2009)
5. Goodwin, C.: Professional Vision. American Anthropologist,. New Series **96**(3) (1994)
6. Charmaz, K.: Constructing Grounded Theory. Sage, Thousand Oaks, CA (2014)
7. Schön, D.A.: The reflective practitioner: how professionals think in action. Basic Books, New York (1983)
8. Döringer, S.: 'The problem-centred expert interview'. Combining qualitative interviewing approaches for investigating implicit expert knowledge. Inter. J. Social Res. Methodol. **24**(3), 265–278 (2021)
9. Dourish, P.: Where the action is: the foundations of embodied interaction. MIT press (2001)
10. Harris, C.G., Achora, J.C.: Designing ICT for Agriculture (ICT4A) Innovations for Smallholder Farmers: The Case of Uganda. In: Proceedings of the XIX International Conference on Human Computer Interaction (Interacción 2018), pp. 1–9. Association for Computing Machinery, New York (2018)
11. Hutchins, E.: Cognition in the Wild. The MIT Press, Cambridge MA (1996)
12. Burchi, G., et al.: Information Technology Controlled Greenhouse: A System Architecture, IEEE IoT Vertical and Topical Summit for Agriculture – Tuscany (IoT Tuscany) (2018)
13. Kirkpatrick, K.: Technologizing agriculture. Commun. ACM **62**(2) (2019)
14. Kuutti, K., Bannon, L.J.: The turn to practice in HCI: towards a research agenda. In: Proceedings of the SIGCHI Conference on Human Factors in Computing Systems (CHI 2014), pp. 3543–3552. Association for Computing Machinery, New York (2014)
15. Lundström, C., Lindblomb, J.: Considering farmers' situated knowledge of using agricultural decision support systems (AgriDSS) to foster farming practices: the case of CropSAT. Agricul. Syst. **159**(C), 9–20 (2018)
16. Rogers, Y., Sharp, H., Preece, J.: Data Gathering. In Interaction Design: Beyond Human Computer Interaction, pp. 273–370. John Wiley and Sons, Hoboken, NJ (2023)
17. Rosala, M., Pernice, K.: User Interviews 101. Nielsen Norman Group. https://www.nngroup.com/articles/user-interviews/, (Accessed 30 Sep 2023)
18. Hussain, S.A.: ICT4Agriculture lessons learned from developing countries. a systematic review protocol. In: Proceedings of the Eighth International Conference on Information and Communication Technologies and Development (ICTD 2016), Article 52, pp. 1–4. Association for Computing Machinery, New York (2016)

19. Zewge, A., Weldemariam, K., Hailemariam, S., Villafiorita, A., Susi, A., Belachew, M.: On the use of goal-oriented methodology for designing agriculture services in developing countries. In: Proceedings of the International Conference on Management of Emergent Digital EcoSystems (MEDES 2012), pp. 40–47. Association for Computing Machinery, New York (2012)

20. Gibson, J.J.: The ecological approach to visual perception. Classic Psychology press, New York (2014)

Implementation Human-Computer Interaction on the Internet of Things Technologies: A Bibliometric Analysis

Andrés Felipe Solis Pino[1,2] , Andrés Felipe Agudelo[1] , Pablo H. Ruiz[2] ,
Alicia Mon[3] , Cesar Alberto Collazos[1] , and Fernando Moreira[4(✉)]

[1] Universidad del Cauca, Calle 5 N.º 4-70, 190003 Popayán, Cauca, Colombia
{afsolis,andresfag,ccollazo}@unicauca.edu.co
[2] Corporación Universitaria Comfacauca - Unicomfacauca, Cl 4 N° 8-30 Centro Histórico,
190001 Popayán, Cauca, Colombia
pruiz@unicomfacauca.edu.co
[3] Universidad Nacional de La Matanza, Florencio Varela 1903 (B1754JEC), 1754 Buenos
Aires, San Justo, Argentina
[4] REMIT, Universidade Portucalense, Porto & IEETA, Universidade de Aveiro, Aveiro, Portugal
fmoreira@upt.pt

Abstract. Human-computer interaction has acquired a preponderant role in the domain of the Internet of Things, facilitating the creation of intuitive interfaces and improving user experiences. Despite this, there needs to be more secondary studies characterizing this domain's current state and conceptual structure, leading to a limited understanding of the field. A bibliometric analysis systematically examined 2106 articles published between 2008 and 2023 from Scopus and Web of Science. Publication metadata was analyzed using Bibliometrix and VOSviewer tools to determine significant trends and patterns in the domain. The study results indicate that the annual growth rate of publications was 41.88%, suggesting a growing interest in the field. China led in productivity and funding, followed by the U.S. and the European Union. The research spanned several disciplines, including computer science, engineering, and mathematics, with key research areas focusing on wearables, gamification, and edge computing. The study's main conclusion relates to this domain's rapid growth and multidisciplinary nature, emphasizing user-centered design to address human needs and provide intuitive interactions. It was found that, as the Internet of Things becomes more widespread, human-computer interaction is set to play a critical role in shaping future Internet of Things environments.

Keywords: Human-computer Interaction · Internet of Things · Bibliometric Analysis · Technology adoption · Scientometrics

1 Introduction

The field of study of Human-Computer Interaction (HCI) is dedicated to designing, evaluating, and implementing computer systems with a focus on their use by humans, dedicated to investigating the phenomena associated with this interaction. Its importance

© The Author(s), under exclusive license to Springer Nature Switzerland AG 2024
Á. Rocha et al. (Eds.): WorldCIST 2024, LNNS 985, pp. 194–203, 2024.
https://doi.org/10.1007/978-3-031-60215-3_19

lies in understanding human-machine interaction's needs, objectives, and patterns. This allows the design and development of technological products that meet these needs and helps to understand people's behaviors, motivations, and goals to improve their interaction with machines [1]. The internet of Things (IoT), a key component of Industry 4.0, has acquired a fundamental role in the daily lives of people and organizations [2]. It is estimated that by 2025, there will be 75 billion interconnected devices performing various tasks, from smart devices that perform tasks in the home to traffic control and environmental monitoring. In parallel, at the enterprise level, the IoT has improved production processes [3].

HCI is essential in operating the IoT because interconnected devices are increasingly integrated into our everyday environment, and interaction with these devices through interfaces allows people to control or configure them [4]. In the context of the IoT, there are challenges and problems associated with HCI. One is interface design, where applying traditional design techniques may result in an awkward coexistence between humans and IoT devices [5]. Another area for improvement is standardization; given the diversity of manufacturers producing IoT devices, uniformity in communication protocols, and user interfaces is often necessary [6]. Finally, there is the challenge of educating the user, as many do not fully understand the capabilities and implications of the IoT, so it is crucial to design interfaces that educate users about the functionalities, and privacy implications [5].

Several primary types of research have been reported regarding the integration of HCI with the IoT, covering a wide range of applications, including home management platforms, emotion recognition systems for this technology [7]. This scenario reflects the growing importance of the scientific community to studies in this field, with multiple pieces of evidence and proposals. In terms of secondary research in the domain of HCI with the IoT, there needs to be more work that defines and characterizes the current state of this field. Turunen et al. in [5] review research on human-IoT interaction, noting that most current applications on these devices use simple graphical interfaces to monitor and configure devices, focusing on initial configuration rather than situated control. In addition, more natural interaction techniques, such as speech and gestures, should be explored. Another review is that of Chen et al. in [8], who indicates that this is a multidisciplinary field encompassing diverse areas such as psychology, and computer science. The main lines of research identified are human-centered design, context-aware computing, and system security. Further secondary studies that characterize the scientific literature regarding the status of the application domain of HCI in the IoT technologies have yet to be documented. Therefore, the present paper addresses this knowledge gap by applying quantitative techniques to the available literature through a bibliometric analysis [9].

The primary distinction of this bibliometric analysis concerning other similar studies is its specific focus on characterizing and determining the conceptual structure of the domain of HCI implementation in IoT technologies. This article aims to conduct a comprehensive bibliometric analysis of the implementation of HCI in the IoT to identify and evaluate the countries, universities, and thematic evolution in this field.

2   Materials and Methods

The framework proposed by Donthu in [10], which provides a set of guidelines for the execution of such studies, was adopted. In addition, elements of the science mapping Workflow methodology [11] were incorporated to facilitate data collection, analysis, and visualization. The main tools employed in this process include the Bibliometrix library [11], which facilitate the comprehensive implementation of bibliometric analysis using text mining techniques, and VOSviewer [12] was used to construct and visualize metadata-based bibliometric networks.

2.1   Search Execution

A search strategy (Table 1) based on the PICOC criteria was applied to collect information in bibliographic databases systematically. In this study, data were obtained from the scientific databases Scopus and Web of Science (WOS), globally recognized in engineering, and considered reliable means for searching literature reviews [13].

Table 1. Search string applied on WOS and Scopus databases

("Human-computer interaction" **OR** "interacción humano-computador" **OR** "Computer-mediated communication" **OR** "Human-machine interaction" **OR** "Human-based computing" **OR** "Human-Technology Interaction" **OR** "Human-computer Interface") **AND** ("internet of things" **OR** "Internet de las Cosas" **OR** "IoT")

In this bibliometric study, a search strategy targeted the titles, abstracts, and keywords of academic articles written in English and Spanish. The inclusion criteria required peer-reviewed articles, ensuring their academic quality and relevance. The search was conducted on the Scopus and Web of Science databases, both renowned for their prestige and long-standing reputation in the scientific community. The search was not confined to a specific timeframe and was executed in July 2023. The search yielded a total of 2262 documents published between 2008 and 2023. Publications with incomplete metadata were excluded from the selection. To avoid redundancy, duplicate studies were identified and removed, leaving only one instance of each in the database. This process identified 313 duplicate studies, reducing the number of unique papers to 2016. Finally, these articles were imported into bibliometrix for further analysis to ascertain the current state of the field of study.

3   Results and Discussion

3.1   General Information on the Domain

To the general information (Fig. 1) derived from the bibliometric analysis applied to the domain of HCI in technologies related to the IoT, a significant period of study spanning from 2008 to 2023 is highlighted, thus comprising 15 years, and providing a longitudinal perspective of the evolution in these areas. A total of 935 sources are identified, including journals, books, and other media, reflecting a considerable diversity of published materials in this domain.

Fig. 1. General information on the domain.

As for the documents found, 2106 were reviewed, which served as the basis for the present study. A noteworthy aspect is the annual growth rate of 41.88%, which indicates a rapid expansion of the domain, denoting a significant interest on the part of the scientific community in integrating these two fields of knowledge; it should also be noted that, compared other IoT domains, such as market analysis, hospitality and tourism, the figures with higher. The average age of the papers is relatively low, being 3.94 years, suggesting that research in this field is mainly recent, with continuous contributions. Along these lines, a relevant indicator is the average number of citations per paper, which is 13.29, reflecting a moderately high impact of the research in the academic community. In addition, the documents referenced by the knowledge base (2106) total 68,238 references, denoting deep bibliographic review and a solid analysis of the scientific and academic body in the documents found. As for the content of the documents, a diversification of keywords plus is observed, reaching a total of 9026 words, revealing a diversity of vocabulary associated with these two fields, highlighting the complexity and thematic diversity of the domain. The author keywords total 5543.

3.2   Annual Scientific Production in the Domain

In this section of the article, the annual evolution of scientific production in the application of HCI on IoT is examined. Figure 2 shows a bar chart illustrating the research output in this field, showing a gradual increase in the number of publications over time, indicating a growing interest in the convergence of these two significant fields by the research community. The trend is upward, with a value of 0.82, corroborating the above; this is in line with other domains of the IoT, where at a general level, an upward trend in publications related to this technology is observed, this could be due to multiple factors, such as its importance in Industry 4.0 and the interconnectivity of society [14]. Finally, the year with the lowest number of publications was 2008, with only one paper, while the year with the highest number was 2019, with 319 papers. In addition, a critical period stands out between 2014 and 2015, with a 100% increase in production from one year to the next.

HCI and IoT research have evolved significantly over the years, reflected in the steady increase of papers published sustainably over time. From its inception in 2008 with a single paper, the field experienced initial growth between 2008 to 2014, marking a phase of exploration and development. The rapid expansion between 2015 and 2019 suggests technological advances and further application of HCI principles on the IoT. However, subsequent years show fluctuations in the number of articles, with notable declines in

2020, 2022, and 2023; these fluctuations can be attributed to factors such as changes in funding, research approaches, and global events, such as the COVID-19 pandemic [15].

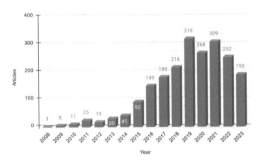

Fig. 2. Annual production of publications in applying HCI on the IoT.

3.3   Universities and Research Centers Are Prominent in the Domain

This section analyzes the top affiliations (Fig. 3) of the most prominent authors in this domain, allowing us to determine the origin of scholarly contributions to the field. With 167 unique affiliations identified worldwide, the Chinese Academy of Sciences emerges as the predominant institution, leading with 31 published papers. Chinese institutions dominate the landscape, such as the Ministry of Education of the People's Republic of China, the University of Chinese Academy of Sciences, and Shanghai Jiao Tong University, among others, making substantial contributions to the domain. Institutions from China, Italy, England, Singapore, and the United States stand out. The Università degli Studi di Bari Al-do Moro and the Politecnico di Milano are among the best Italian universities. The University of Nottingham and University College London also represent the UK.

The World University Rankings indicate that leading institutions, including the National University of Singapore, Tsinghua University, Nottingham University, and the University of Washington, significantly contribute to their respective fields. The QS World University Rankings 2024 further corroborates this, as it lists these academically impactful institutions within the top 1000 contributors to their domains. This underscores the global scientific community's interest in these institutions and their contributions.

The field of HCI on IoT technologies is experiencing an increase in global interest, as evidenced by the 167 affiliations they produce in the domain. Chinese institutions are in the lead, with 8 of the top 10 from this country, indicating a solid focus and substantial funding in this field from the Asian country. This may be due to ambitious Chinese funding programs in emerging technologies such as the IoT, such as the Made in China 2025 program [16]. Similarly, European universities, mainly from Italy and England, continue to contribute significantly, reflecting Europe's continued interest in this area of research.

The presence of prestigious global universities such as the National University of Singapore, Tsinghua University, and the University of Nottingham underlines this field's

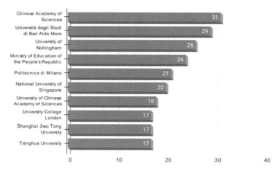

Fig. 3. Institutions highlighted the productivity of applying HCI on the IoT.

importance and emerging nature. Interestingly, there is no clear correlation between the number of articles and the World University Rankings 2023. For example, the Chinese Academy of Sciences appears outside the ranking despite having the highest number of publications. Finally, from a geographical point of view, the contributions span several continents, demonstrating a truly international interest.

3.4 Productivity of Countries in the Domain

This section presents a detailed analysis of the scientific production of HCI applied to the IoT, quantifying the number of publications per author according to their country of affiliation, which allows for determining the scientific contribution of each geographical region to this specific field. Figure 4 shows the research productivity in the domain, with 63 participating countries from different geographical regions. China is the leader, with 1882 publications highlighting its dominant role, aligned with what was found in the most productive university affiliations, where the Asian country also leads. It is followed by the United States, with 807 publications reflecting a significant contribution to the field. Italy, India, and England are among the top five countries, confirming the global distribution of research efforts in the field.

It is important to note that there is a disparity in the frequency of publications, with a contrast between the leading research countries and the nations at the lower end; in this case, 75% of the countries have fewer than 100 publications, indicating the concentrated nature of research productivity. Countries such as Chile, Latvia, Lesotho, Luxembourg, Montenegro, Nepal, Nicaragua, Peru, and Slovenia have the most minor research, with only one publication each. This underscores the need to foster collaborative research and knowledge sharing among nations.

China leads in research productivity in the domain, while the United States and Italy are one step below in production, although with a significant difference. Global interest is not limited to a specific geographic area, as evidenced by the many countries contributing to the field in different regions. Medium-performing countries, such as India, England, Germany, and South Korea, have a high research output but comparatively lower than the leaders. In parallel, many countries contribute on a smaller scale, reflecting a global interest but with a lower output level.

Fig. 4. Productivity segmented by country in the domain.

The data found are skewed to the right, i.e., with high outliers such as China, the United States, and Italy in terms of research productivity, suggesting that research productivity is probably related to factors such as gross domestic product, population, and research expenditure, as reported in another research [17]. A significant difference in research productivity between countries could be funding, infrastructure, number of researchers, and other factors. The Asia-Pacific region has a strong presence in countries such as India, South Korea, and Japan, with high research frequencies in the field. Figure 4 also includes several countries demonstrating diverse interests and involvement worldwide, but their research community may need to be more extensive or focused on other areas or interests.

The top countries, with more than 500 publications after China, are the United States, Italy, India, England, Germany, and South Korea, large developed economies with solid research systems. After this group, medium-sized economies, such as Japan, Australia, and Spain, have between 200 and 500 publications. In general, most countries have some level of research output; however, smaller, less developed nations produce fewer publications in the field. Interestingly, some smaller countries, such as Singapore, Finland, and Denmark, have between 90 to 100 articles pointing to effective research policies in those countries for technologies such as the IoT. It is essential to note that the first seven countries concentrate more than 60% of all publications, indicating that research in this field is concentrated in a few crucial countries.

3.5   Thematic Evolution in the Domain

A segmentation of the thematic evolution of the domain of the application of HCI on IoT is performed using a Sankey chart (Fig. 5). The evolution of HCI within the IoT is a field shaped by the integration of emerging technologies that are incorporated as they mature in parallel with each other. In the early years, from 2008 to 2011, the emphasis was on establishing the basic infrastructure and generating the research corpus, focusing on security, intelligent objects, and the conceptualization of the Web of Things and the IoT. This period marked the initial phase of the technology, remarking the need to address the fundamental principles and conceptual issues of the field, which is a model typically followed by most relatively young domains.

From 2012 to 2018, the field experienced substantial growth and diversification. The emphasis transitioned towards specialized applications and technologies, including but

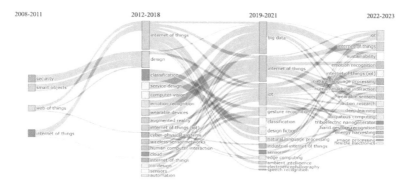

Fig. 5. Thematic evolution of the domain using a Sankey plot.

not limited to service design, computer vision, emotion recognition, wearable technology, augmented reality, cyber-physical systems, wireless sensor networks, and cloud computing. This shift towards specific technologies and applications signifies the field's evolution of intricate systems. It also indicates an increased focus on human factors, user experience, and integrating these technologies into daily life, a current area of challenge and active research.

Between 2019 and 2021, it evolved to include advanced topics such as Big Data, Machine Learning, gesture recognition, classification, fiction design, Natural Language Processing (NLP), Industrial Internet of Things, edge computing, and speech recognition. This phase exposed a trend towards more sophisticated data processing techniques and an expanded range of applications, emphasizing industrial domains. Finally, from 2022 to 2023, there has been a clear focus on sustainability, emotion recognition, NLP, interaction, wearable sensors, deep learning, ubiquitous computing, triboelectric nanogenerators, hand gesture recognition, and image processing. This indicates efforts to create more energy-sustainable solutions, improve user interaction through advanced sensing technologies, and address broader societal implications. The thematic evolution of HCI in the IoT domain is influenced by rapid advances, increasing sophistication, and a growing emphasis on user experience and energy sustainability, from foundational concepts to complex applications integrated into everyday life, reflecting the maturation of the IoT technology with implications for design. Challenges such as sustainability and responsiveness to human needs and emotions are driving innovation and shaping the future landscape of the IoT.

Finally, this bibliometric analysis has some limitations that should be acknowledged. First, the search strategy was limited to two databases and two languages, which may have excluded relevant studies from other sources or languages. Second, the thematic evolution was based on a qualitative interpretation of the keywords and abstracts, which may not capture the full scope and depth of the research topics. Therefore, future research should address these limitations by expanding the search strategy, verifying the metadata, applying more rigorous methods for thematic analysis, and conducting a systematic review or meta-analysis of the studies.

4 Conclusions

This research presents a bibliometric analysis of the application of HCI in the IoT. The bibliographic databases Scopus and WOS were used to carry out this analysis to identify the most significant academic papers in this field. A total of 2106 articles published between 2008 and 2023 were examined. Subsequently, Bibliometrix and VOSviewer tools were used to process the metadata of the collected documents. The main results reveal a constantly developing field with an annual growth rate of 41.88%, with a high level of collaboration between researchers and an increase in annual publications, indicating a growing interest on the part of the research community. In particular, the results point to China as the leading country regarding productivity and funding, followed by the United States and the European Union. In terms of the thematic evolution of the domain, research topics evolved from basic infrastructure and conceptual issues such as smart objects and security to advanced applications in artificial intelligence, virtual reality, edge computing, emotion recognition, and human-robot interaction.

As for future work that can be derived from this work, it is related to secondary studies specifically focused on the principles, methods, and evaluations of HCI design for IoT technologies, which serve as a complementary element to this study. Likewise, in terms of primary research, explicitly addressing the need to develop standardized interfaces and design lines for IoT systems is a priority, as it could unify technological experiences in the field.

Acknowledgements. This work was supported by the FCT – Fundação para a Ciência e a Tecnologia, I.P. [Project UIDB/05105/2020].

References

1. Diederich, S., Brendel, A.B., Morana, S., Kolbe, L.: On the design of and interaction with conversational agents: An organizing and assessing review of human-computer interaction research. J. Assoc. Inf. Syst. **23**, 96–138 (2022). https://doi.org/10.17705/1jais.00724
2. Pivoto, D.G.S., de Almeida, L.F.F., da Righi, R., R, Rodrigues JJPC, Lugli AB, Alberti AM,: Cyber-physical systems architectures for industrial internet of things applications in Industry 4.0: A literature review. J. Manuf. Syst. **58**, 176–192 (2021). https://doi.org/10.1016/j.jmsy.2020.11.017
3. Rudrakar, S., Rughani, P.: IoT based agriculture (Ag-IoT): a detailed study on architecture, security and forensics. Inf Process Agric. (2023). https://doi.org/10.1016/j.inpa.2023.09.002
4. Ferrari, M.I., Aquino, P.T.: Human interaction and user interface design for IoT environments based on communicability. In: Amaba, B. (ed.) Advances in Human Factors, Software, and Systems Engineering, pp. 93–104. Springer International Publishing, Cham (2016). https://doi.org/10.1007/978-3-319-60011-6
5. Turunen, M., Sonntag, D., Engelbrecht, K.-P., Olsson, T., Schnelle-Walka, D., Lucero, A.: Interaction and Humans in Internet of Things. In: Abascal, J., Barbosa, S., Fetter, M., Gross, T., Palanque, P., Winckler, M. (eds.) Human-Computer Interaction – INTERACT 2015, pp. 633–636. Springer International Publishing, Cham (2015). https://doi.org/10.1007/978-3-319-22668-2

 6. Booij, T.M., Chiscop, I., Meeuwissen, E., Moustafa, N., Hartog, F.T.H.D.: ToN_IoT: The Role of Heterogeneity and the Need for Standardization of Features and Attack Types in IoT Network Intrusion Data Sets. IEEE Internet Things J. **9**, 485–496 (2022). https://doi.org/10.1109/JIOT.2021.3085194
 7. Gong, C., Lin, F., An, X.: A novel emotion control system for embedded human-computer interaction in green Iot. IEEE Access **7**, 185148–185156 (2019). https://doi.org/10.1109/ACCESS.2019.2960832
 8. Chen, H., Duffy, V.G.: Systematic Literature Review on Human-Internet of Things (IoT) Interaction in the Era of Ambient Intelligence. In: Duffy, V.G., Ziefle, M., Rau, P.-L.P, Tseng, M.M. (eds.) Human-Automation Interaction: Mobile Computing. Springer International Publishing, Cham, pp 431–451. (2023). https://doi.org/10.1007/978-3-031-10788-7_26
 9. Donthu, N., Kumar, S., Pandey, N., Pandey, N., Mishra, A.: Mapping the electronic word-of-mouth (eWOM) research: a systematic review and bibliometric analysis. J. Bus. Res. **135**, 758–773 (2021). https://doi.org/10.1016/j.jbusres.2021.07.015
10. Donthu, N., Kumar, S., Mukherjee, D., Pandey, N., Lim, W.M.: How to conduct a bibliometric analysis: An overview and guidelines. J. Bus. Res. **133**, 285–296 (2021). https://doi.org/10.1016/j.jbusres.2021.04.070
11. Aria, M., Cuccurullo, C.: Bibliometrix : an R-tool for comprehensive science mapping analysis. J. Informetr. **11**, 959–975 (2017). https://doi.org/10.1016/j.joi.2017.08.007
12. Arruda, H., Silva, E.R., Lessa, M., Proença, D.J., Bartholo, R.: VOSviewer and Bibliometrix. J. Med. Libr. Assoc. JMLA **110**, 392–395 (2022). https://doi.org/10.5195/jmla.2022.1434
13. Pranckutė, R.: Web of Science (WoS) and Scopus: The Titans of Bibliographic Information in Today's Academic World. Publications 9 (2021). https://doi.org/10.3390/publications9010012
14. Sun, X., Wang, X.: Modeling and analyzing the impact of the Internet of Things-based industry 4.0 on circular economy practices for sustainable development: evidence from the food processing industry of China. Front. Psychol. **13** (2022) .. https://doi.org/10.3389/fpsyg.2022.866361
15. Solis Pino, A.F., et al.: Determination of Population Mobility Dynamics in Popayán-Colombia during the COVID-19 Pandemic Using Open Datasets. Int. J. Environ. Res. Public Health **19** (2022). https://doi.org/10.3390/ijerph192214814
16. Hou, J., Li, B.: The evolutionary game for collaborative innovation of the IoT Industry under government leadership in China: an IoT Infrastructure perspective. Sustainability **12**, 3648 (2020). https://doi.org/10.3390/su12093648
17. Meo, S.A., Al Masri, A.A., Usmani, A.M., Memon, A.N., Zaidi, S.Z.: Impact of GDP, spending on R&D, number of universities and scientific journals on research publications among Asian countries. PLoS ONE **8**, e66449 (2013). https://doi.org/10.1371/journal.pone.0066449

Sustainability and Usability Evaluation of E-Commerce Portals

José A. García-Berná[1]([✉]), Sofia Ouhbi[2], Juan M. Carrillo de Gea[1], Joaquín Nicolás[1], and José L. Fernández-Alemán[1]

[1] Department of Computer Science and Systems, University of Murcia, Murcia, Spain
`josealberto.garcia1@um.es`
[2] Department of Infromation Technology, Uppsala University, Uppsala, Sweden

Abstract. Information and communications technologies (ICT) are essential to achieving energy sustainability objectives. With their substantial daily traffic, e-commerce websites are especially important. This study looked into the connection between the most popular e-commerce websites' sustainability and usability. This study evaluated usability employing accepted web-based user experience criteria. Furthermore, a framework for assessing website elements that support sustainability was put forth. Through an analysis of the features of four tools that quantify sustainability-related factors such as page load time, browser cache usage, JavaScript usage, HTTP request volume, and usage of green servers we gathered a total of 39 sustainability criteria for software. Finally, we evaluated whether usability and sustainability were correlated. There was no significant correlation between usability and sustainability according to our research. Still, the framework this paper develops is a useful tool for sustainability evaluations.

Keywords: E-commerce · Software Requirements · Sustainability · Usability

1 Introduction

The availability of a useful website that promotes online sales is essential to the success of e-business [1]. In the context of web development, usability refers to the process of building user-friendly websites. A well-designed website puts the user first and considers a user-centered design, taking into account things like robustness, consistency, response time, visual appeal, and ease of learning [2]. Another important consideration when evaluating a website's quality is its user experience (UX), which sums up how a user feels about a product, service, or gadget after interacting with it [3]. In this study, usability and user experience were employed interchangeably [2].

As e-commerce gains traction, worries regarding its effects on the environment are growing [4]. According to García et al. (2018) [5], sustainability is meeting present needs without endangering the capacity of future generations to meet their own. According to predictions, the computing infrastructure's power consumption will increase to several thousand more TWh by 2030 [6].

Á. Rocha et al. (Eds.): WorldCIST 2024, LNNS 985, pp. 204–213, 2024.
https://doi.org/10.1007/978-3-031-60215-3_20

A senior experience designer claims that the individuals who design websites are accountable for at least 332 million tonnes of CO_2, or 40% of all Internet carbon footprints [7]. This is frequently the result of actions like downloading content that uses a lot of energy, such as high-resolution photos. Power consumption has increased due to the growing number of users and online enterprises, primarily because of energy-intensive servers and less efficient low-end computers.

According to current statistics, almost 33% of people on the planet are consumers. By 2025, this figure is predicted to rise to 2.77 billion[1]. The serious issue of e-commerce website sustainability stems from the demands of both the present and the future. This study investigates possible relationships between sustainability scores and online portal usability. The project aims to promote understanding of the environmental and business aspects of website creation, driven by the idea that *greater usability leads to increased sustainability*. On a number of e-commerce websites, measurements of sustainability and usability will be made with an emphasis on any possible relationships. The selection of e-commerce websites was based on their notable commercial activity and importance in the modern world. Four tools that measure the sustainability qualities of websites were used to construct a framework for evaluating sustainability.

The structure of the paper is as follows. The methodology for evaluating usability and sustainability is described in Sect. 2. Results are depicted in Sect. 3. Finally, a brief discussion and conclusion are given in Sect. 4.

2 Materials and Methods

The section that follows provides a description of every stage in the approach employed for the usability and sustainability assessment. It is important to highlight that the focus of our paper was on e-commerce platforms that are among the most popular websites globally [8].

Usability Evaluation. The most significant e-commerce websites' UX scores were determined using a benchmark provided by the Baymard Institute [9]. The database utilised by this benchmark was derived from a usability analysis of the top 60 B2C retail e-commerce websites. Every assessment considers 702 distinct UX components. For each web portal, a score ranging from 0 to 702 is thus displayed. For this study, these grades have been normalised on a scale of 0 to 10. Because this assessment is updated on a regular basis, the results differ slightly based on when the visit was last made. Table 1 shows the top 60 B2C retail e-commerce sites scores according to the Baymard Institute in UX.

Sustainability Evaluation. We evaluated the energy efficiency of e-commerce portals to assess web design sustainability. We collected a set of metrics from online tools designed for website performance assessment. The sustainability score enabled us to compare the energy efficiency of different e-commerce portals.

---

[1] http://www.oberlo.com/statistics/how-many-people-shop-online.

Table 1. Top 60 B2C retail e-commerce sites according to Baymard

Website	Usab. score (0–10)	Website	Usab. score (0–10)
B&H Photo	9.57	Disney Store	5.87
Home Depot	9.23	Lowe's	5.85
Wayfair	9	Cabela's	5.7
Nordstrom	8.76	Zalando	5.47
Amazon	8.75	Debenhams	5.46
Adidas	7.82	IKEA	5.44
L.L. Bean	7.79	Crate & Barrel	5.4
REI	7.65	Urban Outfitters	5.31
Walmart	7.58	Hayneedle	5.06
ASOS	7.54	American Eagle Outfitters	4.8
Crutchfield	7.48	H&M	4.74
Sears	7.44	Gilt	4.62
Musician's Friend	7.41	Nike	4.62
Target	7.19	Victoria's Secret	4.6
Office Depot	7.09	Newegg	4.46
Neiman Marcus	6.98	Grainger	4.3
Staples	6.89	Overstock	4.1
Best Buy	6.88	Kohl's	4
J.C. Penney	6.84	Ann Taylor	3.92
Build.com	6.79	Williams-Sonoma	3.82
Sephora	6.58	GAP	3.8
Marks & Spencer	6.58	Foot Locker	3.6
Walgreen Co	6.55	Bass Pro Shops	3.59
Etsy	6.45	Argos	3.46
Tesco	6.42	Microsoft	3.19
John Lewis	6.4	HP	2.98
Northern Tool	6.4	TigerDirect	2.91
Macy's	6.24	Costco	2.78
Apple	6.18	Rakuten	2.65
Toys'r'us	6.05	Dell	1.79

Search protocol. Preliminary analysis of major academic databases revealed that web page performance evaluation tools are not extensively explored in academic literature, and their current perspective is limited. These tools are generally aimed at the general public, providing performance metrics. A search was conducted through Google using these search strings webpage efficiency online tool" and "website speed test tool", with only the first page of Google results considered, aligning with the common user behaviour of focusing on the initial search results and potentially abandoning further searches if their desired information is not found.

A set of inclusion criteria (IC) and exclusion criteria (EC) were established in order to choose the tools for performance evaluation: Free (IC1), Web-based (IC2), Provides with performance metrics (IC3), Requires signup (EC1), Not available (EC2), Works on all webpages (EC3), and Exclude CEO oriented tool (EC4).

Tools description. A total of 4 free and web-based tools were selected: *Ecograder, The Green Web Foundation, Website Grader and Google PageSpeed Insights.*

1. **Ecograder** [10] performs the following measurements.
   - **Download speed by searching at Google PageSpeed Insights (PSI) tool.** A higher PSI score indicates greater web efficiency and reduced energy consumption, considering factors like response time, HTML loading speed, and website optimisation.
   - **Average number of HTTP requests.** Minimising HTTP requests overhead, through techniques like browser caching, CSS Sprites, and embedding images in CSS.
   - **Share of resources.** Allowing the browser to access cached resources instead of re-downloading them.
   - **Search capability.** MozRank assesses a website's search capacity based on the quantity and quality of inbound links. Higher-quality inbound links, which consider authority and relevant keywords, result in a better score, indicating the website's significance.
   - **Mobile design.** Responsive web designs ensure seamless performance across various devices, facilitating quick content loading and conserving energy.
   - **Use of flash content.** Avoiding flash animations enhances website usability by reducing energy-intensive downloads and ensuring compatibility with a wider range of devices.
   - **Ecological hosting provider.** This feature considers the fact that servers running websites are powered by renewable energy sources.
2. **The Green Web Foundation** [11] indicates if the web has an ecological accommodation. We analyzed the following aspects:
   - **Cooperativism.** Hosting companies voluntarily provide information about whether their datacenters use renewable energy to the portal.
   - **Checkings.** Periodically, a number of checks are conducted in order to validate some of the claims that the providers may make.
   - **Best-effort.** The Green Web Foundation makes a significant effort to obtain and maintain up-to-date data.
3. **Website Grader** [12]. It offers a score ranging from 0 to 100 to indicate the sustainability of the web. We analysed the following aspects:
   - **Functioning.** Optimizing a website's performance is vital for boosting traffic, conversion rates, customer acquisition, and revenue, constituting 30% of the total score. It is calculated based on page size, the number of HTTP requests, and page speed.
   - **Responsive mobile design.** Taking this aspect into account can attract a larger customer base, as adapting a web tool to various devices promotes sustainability during its usage. This section contributes 30% to the total score.
   - **SEO.** A user-friendly website should have clear, concise page titles, relevant meta descriptions, distinct header tags, and site maps to aid navigation. A total of 30% of the total score depends on these concepts.
   - **Security.** SSL certificates provide visitors with the assurance that a website is trustworthy and authentic while shielding it from attacks. Ten percent of the final score is based on this section.

4. **Google PageSpeed Insights** [13]. PSI assesses various elements, including the time taken for a page to load its initial content, the duration for displaying the entire main page content, the speed of content availability, and the time until the website becomes interactive. We analyzed the following aspects:
   - **Next generation formats**. Formats like JPEG2000, JPEGXR, and WebP offer superior image compression compared to PNG or JPEG formats, resulting in faster downloads and reduced data consumption.
   - **Efficient encoding**. Effectively encoded images use less data and load more quickly.
   - **Appropriate size**. Larger images take longer to load and require more data to download the entire website.
   - **Server's response time**. Navigating pages earlier is made possible by a quick server response time.
   - **Visibility of the text**. A higher score is obtained if the text is still readable while the web source loads.
   - **Image loading strategy**. The web page performs better whenever the loading of images that do not show up on the screen is delayed.
   - **Text compression**. When uncompressed compressible resources are published on a webpage, this feature is taken into account.
   - **Video formats**. Videos are a more effective choice for animated content, especially when compared to large GIFs. To improve web page efficiency, consider using video formats like MPEG4/WebM for animations and WebP for static images.
   - **Browser cache**. There are benefits to using these memories in terms of power consumption.
   - **Block rendering**. Eliminate resources that cause rendering delays, prioritize critical JS/CSS elements for immediate loading, and postpone non-critical JS/styles to enhance web page performance.

Nine categories of software sustainability guidelines were put forth in relation to the aforementioned tools: ecological accommodation, searchability, content, mobile design, operation, multimedia, shared resources, security, and HTTP requests.

3   Results

Sustainability Catalogue. Table 2 and 3 depict a collection of requirements that websites need to fulfil in order to be deemed sustainable. In order to allow for flexibility in sustainability assessments, these sustainability criteria are derived from the four previously mentioned tools and grouped into categories with weighted percentages based on their applicability in the literature. Individual criteria within each group have separate percentages that add up to 100%. This catalogue was created by the authors and verified by three experts.

To calculate the percentage of importance of each group, these steps were followed. First, an initial percentage was assigned for each group to represent

Table 2. Sustainability Catalogue. Part 1. *Acronyms. Ecograder (Ec), Website Garder (WG), Google PageSpeed Insights (PSI), The Green Web Foundation (GWF)*

ID	Guideline	Description	% of Importance	Source
(g1) Ecological accommodation (3.97%)				
R01	Use servers powered by renewable energy		100	Ec, GWF
(g2) Searchability (4.96%)				
R02	Optimize keywords		20	Ec
R03	Use guest blogging	It is the use of blogs to advertise the web	10	Ec
R04	Create content according to the web		20	Ec
R05	Do not keep broken links		20	Ec
R06	Have search capabilities	Having pages linked to the web	30	Ec
(g3) Content (17.85%)				
R07	Enable text compression		25	PSI
R08	Use an appropriate page size	The larger the page, the slower the load will be. For optimal performance, keep the page size below 3MB	20	WG
R09	Write appropriate titles	At most, they must be 70 characters	15	WG
R10	Write appropriate meta descriptions	Relevant and less than 300 characters	15	WG
R11	Use the appropriate labels	Use heading tags to distinguish them from content	15	WG
R12	Have a web map	It is used to navigate faster and easier	10	WG
(g4) Mobile design (0.99%)				
R13	Responsive web design		100	Ec, WG
(g5) Operation (39.29%)				
R14	Time of the first rendering with content	Time it takes to get visual responses from a website. Pages should load in less than 3 s	10	PSI, WG
R15	Time of the first significant rendering	Time it takes to load the main content of a page	10	PSI
R16	Time until it is interactive		10	PSI
R17	Maximum potential FID	The latency of the first interaction that users might experience is the duration of the longest task	5	PSI
R18	Reduced server response time		15	PSI
R19	Minify CSS files	Delete unused ones	5	PSI
R20	Minify JavaScript resources		5	PSI
R21	Reduce JavaScript runtime		10	PSI
R22	Prioritize visible content		10	PSI
R23	Remove the resources blocking the rendering	JavaScript or CSS	10	PSI, WG
R24	Compress JavaScript and CSS	The website will run faster	10	WG

its relative weight in the total grade. In particular, ecological accomodation received a 20%, searchability 5%, content 15%, mobile design 5%, operation 18%, multimedia 10%, shared resources 12%, security 2% and HTTP requests 13%. Then, the importance percentages of each group was normalized. Therefore, to make the distribution of percentages more objective, the following calculations were performed. Results of these calculations are presented in Table 4.

1. The initial percentage of each group g is multiplied by its number of criteria:
   $AbsoluteWeight_g = InitialPercentage_g * CriteriaNumber_g$
2. The sum of all these results is made for the n groups:
   $TotalWeight = \sum_{i=1}^{n} AbsoluteWeight_i$

Table 3. Sustainability Catalogue. Part 2. *Acronyms. Ecograder (Ec), Website Garder (WG), Google PageSpeed Insights (PSI), The Green Web Foundation (GWF)*

ID	Guideline	Description	% of Importance	Source
(g6) Multimedia (9.92%)				
R25	Optimized images	Next generation formats (JPEG 2000, JPEG XR or WebP)	21	Ec, PSI
R26	Suitable size for images		15	PSI
R27	Effective encoding of images		21	PSI
R28	Postpone loading images that do not appear on the screen		22	PSI
R29	Use video formats for animated content	Large GIFs are ineffective at displaying animated content because they use more bytes from the network	21	PSI
(g7) Shared resources (7.14%)				
R30	Use shared hosting account		40	Ec
R31	Use a content distribution network		30	Ec
R32	Store jQuery in Google instead of on your own servers		30	Ec
(g8) Safety (0.40%)				
R33	Secure website	SSL certificates protect websites from attacks	100	WG
(g9) HTTP requests (15.48%)				
R34	Taking advantage of the browser cache		25	Ec, PSI, WG
R35	Avoid multiple page redirects	Page redirects add an additional load cycle, increasing the time to display the page	10	PSI, WG
R36	Number of HTTP requests	Less than 12 is a good number. Between 13 and 59 is average. And 60 or more is high	25	Ec, WG
R37	Use low bandwidth	Avoiding Flash, for example	15	Ec
R38	Speed index	Indicates how quickly the content of a page can be viewed	10	PSI
R39	First CPU idle time	Indicates the first time that the main page thread is inactive enough to receive user action	15	PSI

Table 4. Final percentages for calculating sustainability

Group	Initial %	# Criteria	Absolute Weight	Final %
Ecol. accommodation	20	1	20	3.97
Searchability	5	5	25	4.96
Content	15	6	90	17.85
Mobile design	5	1	5	0.99
Operation	18	11	198	39.29
Multimedia	10	5	50	9.92
Shared resources	12	3	36	7.14
Security	2	1	2	0.4
HTTP requests	13	6	78	15.48
Total	**100**	**39**	**504**	**100**

3. The percentage representing the result obtained from each group g with respect to the total sum is calculated, obtaining the percentage of importance:

$$FinalPercentage_g = \frac{AbsoluteWeight_g}{TotalWeight} * 100$$

The sustainability score of a website was determined by multiplying the score achieved in each criterion by the criterion's percentage of importance and the group's percentage of importance to which the criterion belongs. The final score, on a scale of 0 to 10, was generated by summing all the results. Thus, the contribution of a specific criterion i in group g to the final score was calculated accordingly.

$$ContributionCriterion_i = ScoreCriterion_i * WeightCriterion_i * WeightLevel_g$$

The following formula would be used to determine each criterion's contribution to the final score for a group g of n criteria.

$$ScoreOfTheGroup = WeightOfTheLevel_g * \sum_{i=1}^{n}(ScoreCriterion_i * WeightCriterion_i)$$

Assuming that there are k criteria groups, the final score calculation can be obtained using the following formula.

$$TotalScore = \sum_{i=1}^{k} ScoreOfTheGroup_i$$

The scores for each website are available for download. A Java application that analyses and scores each website was used to determine the final sustainability score[2]. The sustainability scores are shown in the Table 5. A total of 43 sites were given a sustainability score. In some cases it was not possible to calculate the sustainability score of some sites, mostly due to difficulties with the four tools used.

Table 5. Sustainability scores (portals without score are not depicted)

Website	Sustainability Score	Website	Sustainability Score
Office Depot	8.56	Apple	6.46
Williams-Sonoma	8.32	John Lewis	6.43
Microsoft	7.91	Rakuten	6.37
Crutchfield	7.83	Newegg	6.32
B&H Photo	7.59	Sears	6.22
Wayfair	7.49	Debenhams	6.19
GAP	7.29	Walgreen Co	6.06
L.L. Bean	7.03	Marks & Spencer	5.97
Overstock	7.02	Crate & Barrel	5.94
Etsy	6.98	Cabela's	5.89
Victoria's Secret	6.97	Toys'r'us	5.89
TigerDirect	6.94	Nordstrom	5.80
IKEA	6.89	Best Buy	5.67
Dell	6.87	Home Depot	5.60
Costco	6.81	Argos	5.55
Grainger	6.79	Sephora	5.54
Staples	6.76	Amazon	5.16
H&M	6.74	Northern Tool	5.12
Tesco	6.74	Neiman Marcus	4.99
Walmart	6.57	American Eagle Outfitters	4.51
Target	6.50	Ann Taylor	4.20
Disney Store	6.47		

[2] http://umubox.um.es/index.php/s/OY3CuZkAGqAePEt.

4   Discussion and Conclusion

The aim of this study was to investigate the relationship between usability and sustainability in web design, which are key concepts in modern digital societies. This study is in line with SDG7 of the 2030 Agenda. Improving usability contributes to the global transition to sustainable energy sources, and the research looked for ways to increase user satisfaction while reducing energy consumption, which contributes to this goal. Specifically, the study found that Office Depot, Williams-Sonoma and Microsoft's websites of Microsoft scored highly in terms of sustainability, while B&H Photo, Home Depot and Wayfair's websites stood out as having excellent usability. On the other hand, Neiman Marcus, American Eagle Outfitters and Ann Taylor had the least sustainable websites, while Dell, Rakuten and Cosco had the worst usability ratings.

We examined usability and sustainability scores using R programming and R Studio. Sustainability averaged 6.44 ($\sigma = 0.93$), whilst usability averaged 5.85 out of 10 ($\sigma = 1.92$). These averages imply that the analysed websites performed below average in terms of sustainability and usability. A very modest negative correlation ($\rho = -0.0772$) was found using Pearson's correlation coefficient, suggesting a relationship between higher usability and lower sustainability. Remarkably, after looking through more than sixty web pages, no significant correlation was found.

Prior studies have examined the relationship between usability and sustainability, employing Personal Health Records (PHRs). This study produced useful recommendations for sustainable software [14]. Furthermore, it revealed relationships between user interfaces, which may have an impact on the effectiveness and long-term viability of digital platforms. As a result, the relationship between usability and sustainability was confirmed [15].

Enhancing the assessment of usability requires taking into account strategies that go beyond user experience, enabling a more thorough comprehension with a variety of assessment approaches. However, when web portal performance measurements are used for sustainability evaluation, there are certain limits because specific server data access is not available. A thorough assessment of sustainability must include accurate measurements of energy consumption.

This study highlights the connection between sustainability and usability, suggesting for more research into dynamic assessments that take into account actual website usage circumstances. Additional variables such as accessibility and broader webpage categories should be included in a more thorough analysis based on regular user interactions to produce more accurate results. All things considered, this study broadens the comprehension of web developers, auditors, and the digital community regarding the attainment of user-friendly and environmentally conscious online design.

Future work could involve enabling task-specific energy metering, optimising efficiency, and creating user personas to evaluate sustainability and usability for a comprehensive evaluation. Setting up guidelines that explicitly connect usability and sustainability is essential because they give e-commerce platform designers a solid starting point. In addition to supporting sustainable practices in the development of digital platforms, this resource may be essential for designers in achieving more general sustainability objectives.

Acknowledgements. This research is part of the OASSIS-UMU project (PID2021-122554OB-C32) and the Network of Excellence in Software Quality and Sustainability (RED2022-134656-T), all funded by MCIN/ AEI /10.13039/501100011033/ and by ERDF A way to make Europe.

References

1. Venkatesh, V., Agarwal, R.: Turning visitors into customers: A usability-centric perspective on purchase behavior in electronic channels. Manage. Sci. **52**(3), 367–382 (2006)
2. Bevan, N.: What is the difference between the purpose of usability and user experience evaluation methods. In: Proceedings of the Workshop UXEM, vol. 9, pp. 1–4 (2009)
3. Dix, A., Dix, A.J., Finlay, J., Abowd, G.D., Beale, R.: Human-computer interaction, Pearson Education (2003)
4. Mangiaracina, R., Marchet, G., Perotti, S., Tumino, A.: A review of the environmental implications of b2c e-commerce: a logistics perspective. Inter. J. Phys. Distribut. Logistics Manag. (2015)
5. García-Mireles, G.A., Moraga, M.A., García, F., Calero, C., Piattini, M.: Interactions between environmental sustainability goals and software product quality: a mapping study. Inform. Softw. Technol. **95**, 108–129 (2018)
6. Andrae, A.S.G.: Prediction studies of electricity use of global computing in 2030. Int. J. Sci. Eng. Invest. **8**, 27–33 (2019)
7. Christie, J.: Sustainable web design. Application Development, State of the Web (2013). https://alistapart.com/article/sustainable-web-design/, .(Accessed Nov 2023)
8. Top sites ranking for e-commerce and shopping in the world. https://www.similarweb.com/top-websites/category/e-commerce-and-shopping (Accessed Nov 2023)
9. 60 × e-commerce UX case studies. https://baymard.com/ux-benchmark [Accessed Nov 2023]
10. How green is your website. https://ecograder.com (Accessed Nov 2023)
11. The Green Web Foundation. Is your website hosted green? https://www.thegreenwebfoundation.org (Accessed Nov 2023)
12. How strong is your website? https://website.grader.com (Accessed Nov 2023)
13. Pagespeed insignts. https://developers.google.com/speed/pagespeed/insights(Accessed Nov 2023)
14. Garcia-Berna, J.A.: Energy efficiency in software: a case study on sustainability in personal health records. J. Cleaner Product. **282**, 124262 (2021)
15. García-Berná, J.A., Ouhbi, S., Fernández-Alemán, J.L., Carrillo de Gea, J.M., Nicolás, J.: Investigating the impact of usability on energy efficiency of web-based personal health records. J. Med. Syst. **45**(6), 65 (2021)

Who is Really Happier? Re-examining the Portrayal of Happiness on Social Media and the Persistence of Misperception

Asma Elfadl[1]([⊠]), Sameha Alshakhsi[1], Constantina Panourgia[2], and Raian Ali[1]

[1] College of Science and Engineering, Hamad Bin Khalifa University, Doha, Qatar
`aelfadl@hbku.edu.qa`
[2] Faculty of Science and Technology, Bournemouth University, Poole, UK

Abstract. Growing concerns have been raised regarding the potential influence of social media on mental health and well-being, specifically focusing on the phenomenon of social comparison. Prior research has shown that individuals tend to overestimate the happiness portrayed in others' social media posts, resulting in negative outcomes such as low mood, reduced self-esteem, and diminished life satisfaction. However, given the nearly two-decade surge of social media, we question whether this trend persists. This study aims to investigate whether individuals still perceive others' happy posts as happier than their own happy posts on social media, while also exploring potential age and gender differences. Self-reported happiness is a person's perception of their own level of happiness, while perceived happiness is the level of happiness, they believe other people are experiencing. Data was collected via an online survey completed by 314 participants. A mixed ANOVA revealed a significant misperception of happiness, indicating, against the current literature, that individuals tend to overestimate their own happiness compared to the happiness expressed by others in social media posts. Gender emerged as a significant factor influencing happiness misperception, with males reporting higher levels of self-reporting happiness than their happiness. A significant difference between the age groups was found and indicated that the older age group (25–64 years) demonstrated a significantly higher happiness misperception than the emerging adult group (15–24 years). The study reveals new insights on happiness misperception in social media, impacting well-being and social bonding, particularly among males and adults, and altering perceptions of online emotional expressions.

Keywords: Happiness · Perceived happiness · Social media · Misperception · Gender differences

1 Introduction

The pervasive influence of social media in modern society has elicited growing concerns about its effects on mental health and overall well-being [1]. Social media may trigger negative emotions mainly attributable to factors such as peer pressure [2], fear of missing out [3], and low self-esteem [4]. These negative experiences often stem from the

Á. Rocha et al. (Eds.): WorldCIST 2024, LNNS 985, pp. 214–226, 2024.
https://doi.org/10.1007/978-3-031-60215-3_21

fundamental aspect of social comparisons [5], which refer to the tendency of individuals to compare themselves to others. Within the realm of social media, such social comparisons have consistently been associated with various adverse psychological outcomes, including feelings of envy, jealousy, reduced self-image, and poor self-esteem [6].

Social media encourages social comparisons as users are exposed to an abundance of peer-shared images and content. It has been observed that individuals often perceive others' posts on social media as a representative depiction of their actual life, including their level of happiness [7]. However, it is imperative to acknowledge that these perceptions may not accurately reflect reality. The tendency to compare oneself to others on social media can engender feelings of inadequacy and dissatisfaction, as individuals may perceive a discrepancy between their own happiness and the happiness of others portrayed by their social media posts [8]. This highlights the intricate interplay between social media, social comparisons, and the perception of happiness.

Active engagement with social media not only involves the sharing of experiences, but also encompasses the perception and interpretation of others' posts, particularly in terms of happiness. The desire for belongingness and trust in social information influence the enjoyment and happiness derived from these shared experiences [9]. However, the perception of happiness on social media is not solely influenced by external factors, but also by individual characteristics and differences [9].

In the context of social media, there is a distinction between self-reported happiness, which refers to individuals own perceived level of happiness, and perceived happiness, which pertains to the happiness they attribute to others [10]. The filtered and curated nature of social media content can create a distorted reality, where individuals may perceive others as happier than they truly are, contributing to the development of social comparisons [11]. As a result, social media users often believe that others are living happier and more fulfilling lives compared to their own, a phenomenon known as happiness misperception [5]. Happiness misperception in social media posts may play a role in the development of social media addiction or disorder, where individuals excessively use social media for self-validation and continuous comparison with their peers, potentially leading to interference with daily activities and emotional distress [12].

Misperception of happiness lies on how we perceive others' happiness compared to our happiness. Therefore, in order to get a deeper understanding of this concept we need to explore factors involved in social media engagement but also factors affecting perceptions of happiness on social media. The perception of happiness of social media posts is affected by various demographics and personality factors. For instance, individuals with a higher inclination towards fantasy are more likely to emotionally invest in others' lives, including the happiness portrayed in online posts [13]. This heightened emotional involvement can lead to an augmented perception of happiness when exposed to positive content on social media [14].

Among demographics factors influencing happiness misperception, gender-related differences have been observed in their engagement with social media. For instance, some studies have reported that women tend to engage in more frequent social comparisons and experience higher levels of envy on social media platforms as compared to men [15, 16]. Additionally, research has shown that females tend to use social media more than males and are more likely to report negative experiences [17]. Further, women tend to engage in upward social comparison, comparing themselves to those who are better off, while men tend to engage in downward social comparison, comparing themselves to those who are worse off [8]. Consequently, these gender-related tendencies may result in women feeling inferior and experiencing negative emotions when perceiving others as being happier or more successful than themselves.

Of importance, age-related differences also exist in relation to perceptions of happiness and the way individuals engage with social media [18]. In particular, research has demonstrated that younger individuals tend to use social media more frequently and are more inclined to report its negative effects on their mental health [19]. Age-related differences in social media engagement can be attributed to various factors. According to Valkenburg et al. [20], younger individuals tend to possess a stronger need for social validation and acceptance, with social media platforms serving as a mean to fulfil these needs. As a result, younger individuals may be more susceptible to the pressures and influences inherent in social media, leading them to engage in more frequent social comparisons. Furthermore, age may also influence the types of content shared on social media and subsequently the perception of happiness. This may contribute to the perception that younger individuals are happier, as they may be more prone to share content that elicits positive emotions [21].

In previous studies conducted on social media, the primary emphasis revolved around how individuals compare themselves to those who are more prosperous [9, 10]. Recent studies have begun to challenge this perspective by proposing that people are also capable of imitating and adopting the positive emotions displayed by others [22, 23]. The concept of "positive contagion" effect posits that individuals are susceptible to being influenced positively by the emotions and positive experiences depicted in others' social media posts [22, 23]. Moreover, the duration of social media's prevalence for almost two decades warrants a re-evaluation of whether people have become more discerning in assessing the authenticity and reflective nature of others' posts [24].

The current study's goal is to revisit the prevailing assumption that individuals tend to overestimate others' happiness when viewing their seemingly happy posts on social media. In other words, this study aims to explore the current landscape, and examine whether there is a shift in people's perception of happiness on social media. Specifically, our study investigates whether individuals still perceive others' happy posts as happier compared to their own happiness as depicted in their own happy posts. Furthermore, this research explores the relation of the perception and misperception of happiness, on one hand, and age, and gender, on the other. Understanding misperception of happiness on social media, as well as how it is influenced by demographic factors, may provide meaningful suggestions on how social media users should share content online to increase their social capital, promote bonding and encourage social support, consequently improving well-being.

This study, therefore, aims to answer the following research questions:

RQ1: Do happiness in one's own posts and perceived happiness in others' posts on social media vary across gender and age groups?

RQ2: Do users tend to misperceive the happiness portrayed by others in their posts on social media?? Does this misperception apply across different age groups and genders?

We note here that the misperception studied in RQ2 is meant at collective level. A misperception occurs when the majority of users perceive their posts to be happier than others or vice versa.

2 Methodology

2.1 Participants and Procedure

Data were collected via an online survey administered to those who had installed a dedicated application on Google Play, between December 2022 and March 2023. The application is used as a tool to assist users monitor and manage their smartphone usage more effectively. The survey was completed by users who installed the application. The invitation to take part was sent soon after they started using the application. A total of 640 users from various countries responded to the survey and agreed to participate in our study.

Participants who were younger than 15 years old, those who reported not posting on social media and those who reported not viewing others' posts on social media were excluded from the sample. Additionally, participants who failed to correctly answer two attention check questions integrated within the survey were also omitted from the sample. Consequently, only 314 participants met the inclusion criteria and were included in the analytical sample of this study.

Prior to data collection, informed consent was obtained from participants. The study received approval from the Institutional Review Board (IRB) of the first author's institution.

2.2 Measures

Demographic Measures. The participants reported demographic information, including gender and age. Our sample included participants aged between 15 and 64 years.

Happiness About Social Media Posts. Participants' self-reported happiness in their social media posts was measured through the question "Thinking of a relatively happy post about yourself that you recently posted (e.g., achievement you made or a new place you visited), how happy were you in reality?". Their perception of the level of happiness others portray in their social media posts was measured through the question "Thinking of a relatively happy post that your peers recently posted (e.g., achievement they made or a new place they visited), how happy do you think they were in reality?". Responses to these questions were assessed using a ten-point Likert scale, ranging from 1 to 10, with higher scores indicating higher levels of happiness. We have added "I do not post" and "I do not view others posts" as further options.

Data Analysis. The dataset was pre-processed, and the subsequent analysis was conducted using JASP version 0.16.3 [25]. The pre-processing procedure included removing duplicates, handling missing values, removing outliers and creating variables as needed. The data was also transformed and encoded to convert categorical data into numerical form. For this study, participants were divided into two ages groups: emerging adults, defined by the United Nations as individuals aged 15–24 and adults, comprising those aged 25 and above [26]. Discrepancy of happiness variable was calculated by subtracting perceived happiness levels from self- reported happiness levels.

Prior to conducting any test, assumption checks were met, including normality using Q-Q plot and homogeneity of variables as assessed by Levene's test (p > .05). Therefore, parametric tests were employed.

Pearson's correlation analysis was conducted to examine the relationship between variables. A mixed ANOVA test was performed to examine happiness misperception and investigate whether this misperception varies across gender and age. For statistically significant findings, a post-hoc analysis was conducted using Bonferroni's correction. In instances where the results were not statistically significant, a simple main effect analysis was performed to further explore the observed effects. The dataset used in this work can be found in the Open Science Framework[1].

3   Results

3.1   Descriptive Statistics

Descriptive statistics of the participants are presented in Table 1. Of 314 participants, 163 (51.91%) were identified as females and 151 (48.08%) as males. In terms of age group, most participants were aged between 25 and 64 (62.42%), with the remaining participants aged between 15 and 24 (37.58%). Most of the participants reported their country as being the USA (41%) and the UK (22%), followed by France (7%), India and Canada (5% each). The remaining participants were spread in small numbers across the countries.

Table 2 presents the mean values of self-reported and perceived happiness for different gender and age groups. In general, older females reported slightly higher levels of self-reported happiness compared to younger females, but they reported lower levels of perceived happiness. A similar trend was reported by males, with older males reporting higher levels of self-reported happiness compared to younger males. However, the levels of perceived happiness were almost the same for both age groups among males.

We further examined the participants' perceived happiness levels in relation to the actual happiness level. The actual happiness level was determined by calculating the mean value of participants' self-reported happiness levels, yielding an approximate value of 7. Based on the calculated mean value, the perceived happiness levels of participants were divided into three groups: Lower (comprising perceived happiness values less than 7), Similar (categorized perceived happiness values equal to 7), and Higher (categorized perceived happiness values greater than 7). The results are shown in Fig. 1.

Furthermore, we examined the perceived happiness levels in relation to the actual happiness level across gender and age groups, as depicted in Figs. 2 and 3, respectively. The results revealed that majority of males reported perceived happiness levels lower than the actual happiness (49.01%). In contrast, females exhibited a more balanced

---

[1] https://osf.io/gxdzm/?view_only=a5b423f2d9564de4811ed94dc7c37a36.

Table 1. Participants Demographic

Variables	N (314)	%
Gender		
Female	163	51.91
Male	151	48.09
Age		
Emerging Adults (15–24)	118	37.58
Adults (25–64)	196	62.42
Country		
USA	133	42
UK	68	22
France	21	7
Canada	15	5
India	15	5
Australia	14	4
Germany	13	4
Netherlands	11	4
Others	24	8

Table 2. Descriptive Statistics of Self-reported Happiness and Perceived Happiness with Gender Differences

Happiness	Gender	Age	N	M	SD
Self-reported happiness	Female	15–24	48	7.06	1.82
		25–64	115	7.28	1.88
	Male	15–24	70	6.86	2.05
		25–64	81	7.40	1.51
Perceived happiness	Female	15–24	48	7.02	1.95
		25–64	115	6.86	1.80
	Male	15–24	70	6.37	1.83
		25–64	81	6.38	1.69
Discrepancy of happiness	Female	15–24	48	–0.04	1.83
		25–64	115	–0.42	2.18
	Male	15–24	70	–0.49	2.17
		25–64	81	–1.01	1.79

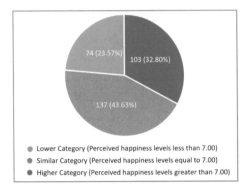

Fig. 1. Perceived Happiness Vs Actual Happiness

distribution between the Lower and Higher categories, with 40.49% reporting perceived happiness levels higher than the actual happiness and 38.65% reporting levels lower than the actual happiness.

Regarding age groups, among those aged 15–24, the majority reported perceived happiness levels lower than the actual happiness (44.07%). Similarly, within the 25–64 age group, the majority of participants reported perceived happiness levels lower than the actual happiness level (43.37%).

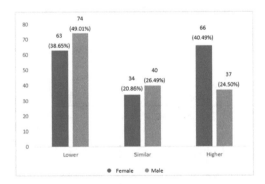

Fig. 2. Perceived Happiness Vs. Actual Happiness Across Gender Groups

3.2 Correlations Amongst Self-reported Happiness, Perceived Happiness, Discrepancy of Happiness, Gender and Age

Pearson's correlation was conducted to examine the relationships between all study variables. The results are presented in Table 3. Self-reported happiness was significantly and positively correlated with perceived happiness ($r = .37$, $p < .001$), suggesting that an increase in self-reported happiness is typically associated with an increase in perceived happiness. Gender showed a significant positive correlation with perceived happiness (r

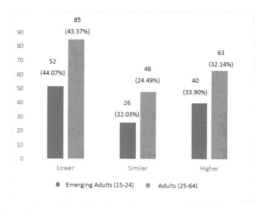

Fig. 3. Perceived Happiness Vs. Actual Happiness Across Age Groups

$= .15, p = .009$), suggesting the females are more likely to have a higher perceived happiness than males. Similarly, the discrepancy of happiness was significantly correlated with gender ($r = -.57, p < .001$).

Table 3. Pearson's correlations between Self-reported happiness, Perceived happiness, Discrepancy of happiness, and Age

Variable	1	2	3	4	5
1. Self-reported happiness	—				
2. Perceived happiness	.37***	—			
3. Age (Emerging adult)	.10	.01	—		
4. Gender (Male)	.02	.15**	.17**	—	
5. Discrepancy of happiness	-.57***	.56***	-.09	.11*	—

* p < .05, ** p < .01, *** p < .001

3.3   Gender and Age Variations in Self-reported and Perceived Happiness

Results from the mixed ANOVA revealed a significant interaction between gender and the discrepancy of happiness [$F(1, 310) = 4.65, p = .032, \eta^2 p = .015$]. However, subsequent post hoc analysis indicated no significant difference between females and males in self-reported happiness. Similarly, there was no significant difference in perceived happiness between females and males.

In relation to age groups, the mixed ANOVA revealed no significant interactions between age groups and both self-reported happiness and perceived happiness. These findings imply that no difference was found between age and individuals' self-reported or perceived levels of happiness.

3.4 Happiness Misperception in the Sample as a Whole

The mixed ANOVA results revealed a significant within-subject main effect for happiness misperception, [$F(1, 310) = 16.51$, $p < .001$, $\eta^2 p = .05$], with $\eta^2 p$ indicating small to moderate effect size. The same results were confirmed when conducting the post hoc analysis, which indicated that self-reported happiness ($M = 7.18$, $SD = 1.82$) was significantly higher than perceived happiness ($M = 6.65$, $SD = 1.81$) in the total sample.

3.5 Happiness Misperception Across Gender Groups

The mixed ANOVA results also revealed a significant interaction effect between happiness misperception and gender. The follow-up post hoc analysis indicated no significant difference when comparing females' self-reported happiness ($M = 7.22$, $SD = 1.86$) to their perceived happiness ($M = 6.91$, $SD = 1.84$), suggesting no evidence of misperception among the female group. In contrast, a significant difference emerged when comparing males' self-reported happiness ($M = 7.15$, $SD = 1.79$) to their perceived happiness ($M = 6.38$, $SD = 1.75$). This finding suggests misperception of happiness among males who reported higher levels of self-reported happiness than perceived happiness of others' social media posts with small effect size of 0.41 (Table 4).

Table 4. Post hoc analysis for the happiness Misperception and Gender

		M Difference	SE	t	Cohen's d	p bonf
Self-reported	Perceived	.49	.12	4.06	.27	< .001***
Female, Self-reported	Male, Self-reported	.04	.22	0.21	.02	1.000
	Female, Percived	.23	.18	1.31	.13	1.000
	Male, Percived	.79	.22	3.70	.44	.001**
Male, Self-reported	Female, Percived	.19	.22	0.87	.10	1.000
	Male, Percived	.75	.17	4.52	.41	< .001***
Female, Percived	Male, Percived	.56	.22	2.63	.31	.053

* $p < .05$, ** $p < .01$, *** $p < .001$

3.6 Happiness Misperception Across Age Groups

In relation to age groups, the mixed ANOVA demonstrated no significant interaction between happiness misperception and age, [$F(1, 310) = 3.51$, $p = .062$, $\eta^2 p = .01$], indicating no significant difference in happiness misperception across age groups. The subsequent simple main effect analysis showed a significant difference in the age groups, specifically within the adult group (25–64).

3.7 Happiness Misperception Across Age and Gender

The mixed ANOVA also demonstrated no significant interaction between happiness misperception, age, and gender (F = .10, p = .75). The subsequent simple main effect showed that for the 15–24 age group, there was no significant happiness misperception among females (F = .03, p = .876) or amongst the male group (F = 3.52, p = .065). These findings indicate that within this age group, gender did not significantly influence happiness misperception. However, in the 25–64 age group, a significant happiness misperception was found among females (F = 4.21, p = .042) and also amongst males (F = 25.84, p < .001).

4 Discussion

This study contributes to the literature by exploring the relationship between self-reported happiness and perceived happiness in social media posts, considering the impact of demographic factors like age and gender. Contrary to previous research [5, 10, 11], participants did not overestimate happiness in others' social media posts compared to their own. This finding challenges the common notion that exposure to happy social media content leads to envy and poor psychological well-being [1, 6]. The study suggests a shift in perception, possibly because individuals may engage in a more critical processing of information on social media, actively questioning the accuracy and authenticity of the content they encounter [27]. Also, skepticism towards online portrayals of happiness [28], and emotion regulation in digital contexts [29].

In relation to happiness misperception across gender differences, the findings of this study showed that female participants did not perceive their own posts as significantly happier than their self-reported happiness. This is surprising as it contrasts previous findings from literature showing that women tend to engage more in self-presentation and selective sharing of positive aspects of their lives on social media [30, 31]. Based on existing evidence, we would expect women to perceive others' happier than them, however this was not the case. While women have been shown to emphasize positive aspects in their online self-presentation, various factors can influence the perception of happiness in their posts. These factors include the content consumed, the range of individuals followed, and the dynamics of social comparison processes [30, 31] and poor psychological well-being [1].

Males were found to report themselves happier than others on social media posts. Males' focus on communication and pragmatic use of social media may elucidate these results. Previous studies have highlighted men's tendency to view social media as a mean of practical communication rather than self-presentation [30, 31]. Therefore, men's self-reported happiness in their social media posts may indicate a willingness to conform to prevailing social norms and establish a positive online image.

Additionally, this study's findings showed that no difference in misperception across different age groups. However, in the 25–64 age group, a significant happiness misperception was found among females and amongst males as well. This may reflect how generational aspects and societal expectations may shape the perception of happiness on social media among older individuals.

The study's findings underline the disparity between self-reported and perceived happiness on social media, highlighting the need for authenticity in online interactions [21]. It encourages genuine self-expression and sharing a broader range of emotions to foster healthier online relationships and reduce the impact of negative social comparisons [9, 21]. Misperceptions of happiness also affect the concept of positive contagion in social networks, emphasizing the importance of accurate perception for positive emotional communication [32, 33].

Despite the above important implications, the current investigation presented some limitations. We should note that the sample may not represent the broader population as the majority of the participants were from the US, followed by UK. Also, data collection was restricted to individuals who had installed an app on Google play introducing selection bias. A further limitation arises from the reliance on self-reported measures and thus the possibility of common sources bias must be acknowledged. Exploring the mechanisms underlying this shift in misperception of happiness as depicted on social media posts would provide us with more meaningful suggestions in the future about the use of social media posts for the promotion of well-being.

In spite of these limitations, it is important not to lose sight of this study's strengths as it offers new insights into the changing trend of misperceiving happiness on social media. While previous studies have highlighted the tendency for individuals to overestimate others' lives online [34] the current study reveals a shift in this pattern. It proposes that individuals are becoming more discerning in their evaluations and no longer perceive others as happier than themselves solely based on the positive content they encounter. Therefore, the misperception of happiness on social media raises ethical concerns, emphasizing the need for social media platforms to prioritize user well-being over engagement metrics. Implementing transparency measures, such as labelling or content moderation strategies, can help users distinguish between genuine expressions of happiness and overly idealized portrayals [23]. Promoting authenticity and reducing the pressure to constantly present a facade of happiness can contribute to a healthier online environment.

5   Conclusion

This study underscores the importance on reflecting on happiness misperception in social media and actively promoting authenticity in content creation and sharing. By prioritizing genuine self-expressions and reducing pressure to present a façade of constant happiness can lead to a healthier and more supportive online environment. Further research is needed to explore additional factors that may influence the perception of happiness in social media posts, such as personality traits, cultural differences, and the authenticity of content. Understanding these factors can contribute to the development of interventions and strategies to promote authentic and healthy social media use.

Acknowledgement. This publication was made possible by NPRP 14 Cluster Grant number NPRP 14C-0916-210015 from the Qatar National Research Fund (a member of Qatar Foundation). The findings herein reflect the work and are solely the responsibility of the authors.

References

1. Kross, E., et al.: Facebook Use predicts declines in subjective well-being in young adults. PLoS ONE **8**(8), e69841 (2013). https://doi.org/10.1371/JOURNAL.PONE.0069841
2. Chin-Hooi, P., Wai, K., Yeik, K., Hwa, P., C-h, P.: Parents vs peers' influence on teenagers' Internet addiction and risky online activities. Telemat. Inform. **35**(1), 225–236 (2018). https://doi.org/10.1016/j.tele.2017.11.003
3. Alutaybi, A., Al-Thani, D., McAlaney, J., Ali, R.: Combating fear of missing out (FoMO) on social media: the FoMO-R method. Int. J. Environ. Res. Public Health **17**(17), 1–28 (2020). https://doi.org/10.3390/IJERPH17176128
4. Zeng, G., et al.: Problematic internet usage and self-esteem in Chinese undergraduate students: the mediation effects of individual affect and relationship satisfaction. Int. J. Environ. Res. Public Health **18**(13), 6949 (2021). https://doi.org/10.3390/IJERPH18136949
5. Samra, A., Warburton, W.A., Collins, A.M.: Social comparisons: a potential mechanism linking problematic social media use with depression. J. Behav. Addict. **11**(2), 607–614 (2022). https://doi.org/10.1556/2006.2022.00023
6. Abdellatif, M.: The impact of social media on life satisfaction: the mediating role of social comparison, envy and self-esteem. researchgate.net (2022). https://doi.org/10.18576/isl/110536
7. Chae, J.: Reexamining the relationship between social media and happiness: the effects of various social media platforms on reconceptualized happiness. Telemat. Informatics **35**(6), 1656–1664 (2018). https://doi.org/10.1016/J.TELE.2018.04.011
8. Vogel, E.A., Rose, J.P., Roberts, L.R., Eckles, K.: Social comparison, social media, and self-esteem. Psychol. Pop. Media Cult. **3**(4), 206–222 (2014). https://doi.org/10.1037/PPM0000047
9. Vogel, E.A., Rose, J.P., Okdie, B.M., Eckles, K., Franz, B.: Who compares and despairs? The effect of social comparison orientation on social media use and its outcomes. Pers. Individ. Dif. **86**, 249–256 (2015). https://doi.org/10.1016/J.PAID.2015.06.026
10. Chou, H.T.G., Edge, N.: 'They are happier and having better lives than i am': the impact of using Facebook on perceptions of others' lives. Cyberpsychol. Behav. Soc. Netw. **15**(2), 117–121 (2012). https://doi.org/10.1089/CYBER.2011.0324
11. Hou, Y., Xiong, D., Jiang, T., Song, L., Wang, Q.: Social media addiction: its impact, mediation, and intervention. Cyberpsychology J. Psychosoc. Res. Cybersp. **13**(1), 4 (2019). https://doi.org/10.5817/CP2019-1-4
12. Beyari, H.: The relationship between social media and the increase in mental health problems. Int. J. Environ. Res. Public Health **20**(3), 2383 (2023). https://doi.org/10.3390/ijerph20032383
13. Himichi, T., et al.: Development of a Japanese version of the interpersonal reactivity index. Shinrigaku Kenkyu **88**(1), 61–71 (2017). https://doi.org/10.4992/JJPSY.88.15218
14. Shiota, S., Nomura, M.: Role of fantasy in emotional clarity and emotional regulation in empathy: a preliminary study. Front. Psychol. **13** (2022). https://doi.org/10.3389/FPSYG.2022.912165/FULL
15. Haferkamp, N., Krämer, N.C.: Social comparison 2.0: examining the effects of online profiles on social-networking sites. Cyberpsychol. Behav. Soc. Netw. **14**(5), 309–314 (2011). https://doi.org/10.1089/CYBER.2010.0120
16. Krasnova, H., Spiekermann, S., Koroleva, K., Hildebrand, T.: Online social networks: why we disclose. J. Inform. Technol. **25**(2), 109–125 (2010). https://doi.org/10.1057/jit.2010.6
17. Perloff, R.M.: Social media effects on young women's body image concerns: theoretical perspectives and an agenda for research. Sex Roles **71**(11–12), 363–377 (2014). https://doi.org/10.1007/S11199-014-0384-6

18. Primack, B.A., et al.: Social media use and perceived social isolation among young adults in the U.S. Am. J. Prev. Med. **53**(1), 1–8 (2017). https://doi.org/10.1016/J.AMEPRE.2017.01.010

19. Abi-Jaoude, E., Naylor, K.T., Pignatiello, A.: Smartphones, social media use and youth mental health. CMAJ **192**(6), E136–E141 (2020). https://doi.org/10.1503/CMAJ.190434

20. Valkenburg, P.M., Peter, J., Schouten, A.P.: Friend networking sites and their relationship to adolescents' well-being and social self-esteem. Cyberpsychol. Behav. **9**(5), 584–590 (2006). https://doi.org/10.1089/CPB.2006.9.584

21. Bailey, E.R., Matz, S.C., Youyou, W., Iyengar, S.S., Matz, S.C.: Authentic self-expression on social media is associated with greater subjective well-being (2020).https://doi.org/10.1038/s41467-020-18539-w

22. Goldenberg, A., Gross, J.J.: Digital emotion contagion. Trends Cogn. Sci. **24**(4), 316–328 (2020). https://doi.org/10.1016/J.TICS.2020.01.009

23. Kramer, A.D.I., Guillory, J.E., Hancock, J.T.: Experimental evidence of massive-scale emotional contagion through social networks. Proc. Natl. Acad. Sci. U. S. A. **111**(24), 8788–8790 (2014). https://doi.org/10.1073/PNAS.1320040111/ASSET/E6DFA794-CF07-443A-8FD7-0DFD58FEF380/ASSETS/GRAPHIC/PNAS.1320040111FIG01.JPEG

24. Davenport, S.W., Bergman, S.M., Bergman, J.Z., Fearrington, M.E.: Twitter versus Facebook: Exploring the role of narcissism in the motives and usage of different social media platforms. Comput. Hum. Behav. **32**, 212–220 (2014). https://doi.org/10.1016/J.CHB.2013.12.011

25. "JASP - A Fresh Way to Do Statistics," (2018). https://jasp-stats.org/. Accessed 03 June 2023

26. "Youth matters: equipping vulnerable young people with literacy and life skills - UNESCO Digital Library," UNESCO institute for lifelong learning, Nov. (2013). https://unesdoc.unesco.org/ark:/48223/pf0000223022. Accessed 08 June 2023

27. große Deters, F., Mehl, M.R.: Does posting Facebook status updates increase or decrease loneliness? An online social networking experiment. Soc. Psychol. Pers. Sci. **4**(5), 579–586 (2013). https://doi.org/10.1177/1948550612469233

28. Bermúdez, J.P.: Social media and self-control: the vices and virtues of attention (2017)

29. Verma, A., Islam, S., Moghaddam, V., Anwar, A.: Encouraging emotion regulation in social media conversations through self-reflection. March (2023)

30. Haferkamp, N., Eimler, S.C., Papadakis, A.M., Kruck, J.V.: Men are from Mars, women are from Venus? Examining gender differences in self-presentation on social networking sites. Cyberpsychol. Behav. Soc. Netw. **15**(2), 91–98 (2012). https://doi.org/10.1089/CYBER.2011.0151

31. Fogg, B.J., Soohoo, C., Danielson, D.R., Marable, L., Stanford, J., Tauber, E.R.: How do users evaluate the credibility of web sites?: a study with over 2,500 participants (2003). https://doi.org/10.1145/997078.997097

32. Christakis, N.A., Fowler, J.H.: Social contagion theory: examining dynamic social networks and human behavior. Stat. Med. **32**(4), 556–577 (2013). https://doi.org/10.1002/SIM.5408

33. Herrando, C., Constantinides, E.: Emotional contagion: a brief overview and future directions. Front. Psychol. **12**, 712606 (2021). https://doi.org/10.3389/FPSYG.2021.712606/BIBTEX

34. Zhong, J., Wu, W., Zhao, F.: The impact of Internet use on the subjective well-being of Chinese residents: from a multi-dimensional perspective. Front. Psychol. **13**, 950287 (2022). https://doi.org/10.3389/FPSYG.2022.950287

Author Index

Printed in the United States
by Baker & Taylor Publisher Services